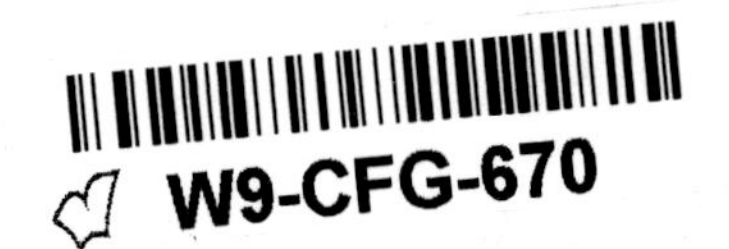

Cranial Creations in Physical Science

Interdisciplinary and Cooperative Activities

Charles R. Downing and Candace J. Aguirre

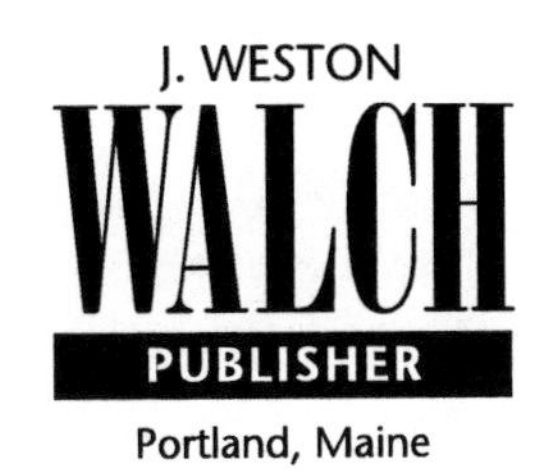

User's Guide
to
Walch Reproducible Books

As part of our general effort to provide educational materials which are as practical and economical as possible, we have designated this publication a "reproducible book." The designation means that purchase of the book includes purchase of the right to limited reproduction of all pages on which this symbol appears:

Here is the basic Walch policy: We grant to individual purchasers of this book the right to make sufficient copies of reproducible pages for use by all students of a single teacher. This permission is limited to a single teacher, and does not apply to entire schools or school systems, so institutions purchasing the book should pass the permission on to a single teacher. Copying of the book or its parts for resale is prohibited.

Any questions regarding this policy or requests to purchase further reproduction rights should be addressed to:

Permissions Editor
J. Weston Walch, Publisher
321 Valley Street • P. O. Box 658
Portland, Maine 04104-0658

—J. Weston Walch, Publisher

1 2 3 4 5 6 7 8 9 10

ISBN 0-8251-2547-2

Printed in the United States of America

Contents

Reproducible Student Pages

Creations—Organized by Subject Area

Reproducible Student Pages

Creations—Organized by Type of Exercise

Area/Topic	Title	Descriptor
Analysis: 9 exercises		
1. General Science	Are You What You Eat?	Determining nutrient percentages
3. General Science	Energy: The Choice Is Yours	Ranking energy sources
7. Meteorological Science	Rain, Rain, Go Away!	Analyzing large amounts of data
11. Earth Science	My Fossil's Older Than Your Fossil	Determining ancestral progression
14. Earth Science	Sounds Pretty Deep to Me	Drawing profiles from sounding data
18. Space Science	Problems in Space Travel	Space travel problems/solutions
22. Space Science	Lost in Space	Determining location from data
23. Space Science	How Many *Elis* Tall Are You?	Exercises in equivalency conversions
27. Physical Science	One Step at a Time	Determining the correct sequence
Art: 6 exercises		
13. Geologic time	"Time" for a Vacation Trip	Travel poster
15. Any earth science	Move Over, Gary Larson	Drawing a cartoon
17. Atmospheric science	Spend Some Time in the "O" Zone	Ozone poster
24. Solar system	A Trip to My Favorite Planet	Travel poster
25. Solar system	"The Creature That Came from . . ."	Design a creature for a planet
34. Any physical science	Move Over, Bill Watterson	Drawing a cartoon
Creative Writing: 12 exercises		
4. General Science	Don't Get Bogged Down on This	Writing about a historical period
6. General Science	My Life as a Fish	Step-by-step writing assignment
9. Meteorological Science	My Life as a Cloud	A story about how clouds form
10. Earth Science	Let's Get Mineralized	Describing how fossils form
12. Earth Science	There'll Be a Hot Time in the Old . . .	What happens when lava flows
16. Earth Science	Rock On!	The process of rock formation
19. Space Science	Death of the Sun	What might happen when sun dies
20. Space Science	An Interview with Galileo	Dialogue with the famous astronomer
21. Space Science	Story of a Star	A story of how stars form and age
26. Physical Science	That Really Burns Me Up	Dialogue between skin and brain
32. Physical Science	Eddie (or Edie) Electron	Description of current flow in circuit
35. Physical Science	Move It!	Description of key terms in motion
Critical Writing: 2 exercises		
2. General Science	Heavy-duty Writing Assignment	How to do a short research paper
5. General Science	Science in (and as) History	How a scientist fits in history
Laboratory Skills: 6 exercises		
8. Meteorological Science	Fluffy as a Cloud	Determining the dew point
28. Physical Science	Liver . . . Onions in Peroxide Sauce . . .	Exothermic reaction
29. Physical Science	Funnels of Fun	Determining the limiting factor
30. Physical Science	Sometimes You Feel Like a Nut	Calories in nuts
31. Physical Science	Faster Than a Speeding . . . Snail?	Using a snail for simple physics
33. Physical Science	Fish Respiration . . .	Dissolved gases

Foreword

Thanks for purchasing this, the second volume of *Cranial Creations.* Owen Miller, the co-author of the first volume, has retired. I have been most fortunate to collaborate on this book with an outstanding and very creative teacher, Candace Aguirre. We have known each other longer than either cares to admit. We enjoy working together nearly all of the time, and had some fun inventing these exercises.

Candace and I have tried to take some of the topics in the earth and physical sciences and apply a creative, and perhaps a bit eccentric approach to their instruction. We have tested these assignments in our own classes. They do work with kids.

If you are like us, you probably spend time modifying many activities to suit your style of teaching. While we would like to think that there is enough assignment variety in this book to appeal to all teachers, we know that there are places where we did not do as well as you would have done. For those times, please modify, mold, and most of all *create.*

If you use only these exercises, we will have been glad to help.

If you develop similar or more bizarre assignments on your own, and these helped inspire you, we will really feel good.

We would like to hear from you, good or bad. Feel free to write us at

Monte Vista High School
3230 Sweetwater Springs Blvd.
Spring Valley, CA 91977.

Chuck Downing
Candace Aguirre
Professional Science Educators

Teaching Guide

Procedural Suggestions for a Typical *Cranial Creations* Creative Writing Exercise

We devised this suggested set of instructions for the creative writing exercises, but these basic directions could be modified to fit other types of exercises. In fact, you will probably notice similarities to some of the analysis-type exercise instructions.

Each student will do some form of assignment individually, either in class or at home. It might range from something as uncomplicated as answering a question in writing to some of the more elaborate homework projects presented in this book. This individual work will be called the "story" throughout the remaining steps.

When the Story is Turned In

1. Collect the "story" papers and stamp them, or mark them in some other way, to show that the individual work was completed, or at least attempted.
2. Hand out the In-Class Student Follow-up questions to be answered individually while you are checking the story papers.
3. If the story was homework, students who did not do the story do not receive these individual questions. Their job is to complete the story in class while the other students do steps 4–9. You need to decide whether to count this makeup work for the same credit as for those who did the homework.
4. Form groups of three (or four, maximum) students. Have each group elect or appoint a scribe who will write down the consensus answers.
5. Hand the stories back to their authors.
6. Each student reads his or her story to the others in the group.
7. After each story is read, the group makes a positive comment about the story to the author (e.g., "Good job!").

8. After listening to all of the stories, the group must come to a **consensus** on answers to the individual questions. These answers are written by the scribe on the consensus paper (see sample on page *xi*). The sample page has space for six answers because it is our experience that a hardworking group of students will be able to come to a legitimate consensus on five to seven questions in a class period, if they also do the preceding steps.

9. Each group member must sign the consensus paper, indicating agreement with the consensus answers. The scribe then gives you the consensus paper. Individual papers may be stapled to the consensus paper if you want to see all of the group work.

Each person in the group gets the same grade for this group portion of the work. It is based exclusively on:

1. doing the story to begin with.

2. the quality and consistency of the consensus answers.

While the group work is being done, we recommend that you follow some form of structured guidelines for group conduct. Dr. Harry Wong has published a set of four such guidelines that have proved a very valuable tool in our group work. If you do not have an established code of conduct for group work in your classroom, we recommend using Dr. Wong's four rules:

1. You are responsible for your own work and behavior.

2. You must ask each support buddy (group member) for help if you have a question.

3. You must be willing to help any support buddy (group member) who asks for help.

4. You may ask help from the teacher only when the group has the same question.

from *Successful Teaching* by Dr. Harry K. Wong

Using these rules will simplify your supervision of group work because you will only be going to groups who have a legitimate question that none of the group members can answer. To check and see if all group members were asked the question before you were called in, ask a group member who did not have a hand raised what the question is. If that group member does not know the question, refuse to answer and leave the group. As soon as your students find out that you are serious about only answering group questions, you will find far fewer hands raised. Following these rules makes group work much more rewarding for both you and the students.

Scribe's Name ______________________ Date ______________

Consensus Answer Sheet

Name of Exercise: ______________________________

You must agree on one **consensus** answer for each question you answered individually. Place those answers next to the numbers below.

1.

2.

3.

4.

5.

6.

Now, sign below in the proper space if you agree with the entire consensus, or indicate which answer(s) you disagree with and why.

Name:	Except This Number:	Because:
______________	________	______________________
______________	________	______________________
______________	________	______________________
______________	________	______________________

Teaching Notes and Answers to Exercises

General Science: Analysis

1. Are You What You Eat?

Teacher Instructions

A few days before you do this exercise, tell the students they each must bring in three food labels for that day. Realistically, the due date for the labels should be the day before you plan to do the analysis, as this will allow you to encourage those students without labels to bring them the next day.

You might want to make a collection of labels without the percentages on them and keep them as a class set. When the new label legislation goes completely into effect, some of this analysis will already have been done on the label.

Groups of three students who each have three labels will provide an excellent pool of information to evaluate.

If you use this as an introductory lesson to some type of nutrition unit, you can "play dumb" about what is natural or artificial and force the students to come to a consensus about what is or is not natural.

If the assignment is more of a closure exercise for a nutrition unit, the students should already have a good idea about natural and artificial ingredients.

If you use the assignment as a way to incorporate some math skills more than as a way to provide information on nutrition, you can accept student decisions on natural vs. artificial or take the time to explain during the period.

The instructions imply that this is entirely a group assignment, but the questions can be answered individually or as a group. The choice is yours.

Answers to Questions

All of the answers will depend on the actual labels used by the students.

General Science: Critical Writing

2. Heavy-duty Writing Assignment

Researching a Specific Piece of Science

Teacher Instructions

This assignment is designed to allow students in any science discipline to do some library research, follow a structured time line for production of a written report, and produce a report of some quality.

The time line is left up to you. Depending on the age and ability of your classes, somewhere between three and six weeks should be adequate for the research and writing of the report. A sample time line for a heterogeneous group of 10th-grade science students is:

Step 1: three days after assignment.
Step 2: one week after assignment.
Step 3: two weeks after assignment.
Step 4: four weeks after assignment.
Step 5: six weeks after assignment.

If you have a bunch of old science magazines stacked on a shelf, you might want to allow student access to them in Step 1. We keep a file of *Scientific American* articles alphabetized in a file cabinet for student use. We cut the articles out of the magazines and staple them into manila folders. Students check them out on a sign-out sheet. They return the article(s) they checked out with their reports.

For grading purposes, steps 1, 2, and 3 are just check points. We usually require submission of a cover sheet that has places for teacher approval of each of the steps on it. A day or two after the topic has been submitted, we check the original article, then we confer briefly with each student about his or her topic while the class is engaged in some form of desk work.

Students will want to know if they can change their topic. We always say, "Yes, providing the cover sheet reflects our approval of the change of topic."

When checking the preliminary bibliography, be sure to make any notations regarding your preference for *real* science magazines as opposed to *Glamour, Hot Rod,* and *Seventeen.* It is advisable to post a list of accepted periodicals prior to the first step. This will alleviate many moments of trauma for both the students and yourself.

If you would like a more concrete sample bibliography than the one in the general student instructions, you can copy and hand out the one that follows the instructions for this exercise. Oh, yes, not *all* of the entries are actual publications.

For a more complete report, you might want to add the concept of referencing to your students. If this is the case, the paragraph at the end of the following sample bibliography should be photocopied and assigned for reading along with the other instructions and sample bibliography.

Scribe's Name ______________________________ Date ____________________

Sample Bibliography

Magazines

Bilical, U.M. "Navel Contemplation." Journal of Meditation and Trances, Vol. 3 (April 1985), pp. 615–690.

A very complete look at the concept and techniques of contemplation of the human navel. Included are sections on correct neck-bending angles, how to unfold from the lotus position, and what to do if lint appears in the navel during contemplation.

Clotron, Cy. Dr. "Should the Police Have the Authority to Give Speeding Tickets to Electrons?" Journal of Atomic Division, Vol. 1, No. 1 (January 1971), pp. 7–9.

This article deals with the question of legal jurisdiction in speeding cases involving fast-moving electrons. The debate between local, state, and federal authorities is given extensive coverage. The range of monetary fines imposed throughout the country is presented in table form.

Dicot, Woody. "Support of the Stem." Xylemtific American, Vol. 15, No. 6 (June 1988), pp. 250–258.

Cellulose as a supporting compound is the focus of this article. Construction of the cell wall is investigated. Movement of water and minerals from root to leaf is presented in clearly labeled diagrams.

Frog, Kermit E. "Exercise to Increase Your Frog's Jump." Croaker's Digest, Vol. 17 (Jan. 1987), pp. 6, 8, 10, 12.

Aerobic exercises as well as power-lifting techniques—all designed to strengthen your amphibian's hind legs—are presented. Step-by-step instructions for both the aerobic and power exercises include photographs of muscular members of the genus Rana.

Pigina, Pit. "Swine as Luau Guests." National Ge-HOG-raphic, Vol. 4 (April 1, 1981), pp. 1–42.

A complete menu of swine favorites and methods of producing the entire menu without offending too many of the guests is presented in this article. How to deal with rioting boars is given only cursory coverage.

(continued)

Scribe's Name ______________________________ Date ____________________

Sample Bibliography *(continued)*

Books

Claws, Sandy. Lions of the Sahara Desert. Cairo, Egypt: Yule-Tide Publishing, 1988.

A daily journal of a scientist's observations in the Sahara Desert is presented in narrative form. Chapters on feeding habits and life in the pride are especially informative.

Crawford, T. James, et. al. Century 21 Keyboarding, Formatting, and Document Processing. Cincinnati: South-Western Publishing Co., 1987.

This text provides the basics of keyboarding, formatting, and document processing in an understandable fashion. The text is written at a level appropriate for report producers of a variety of skill levels.

Gibaldi, Joseph, and Walter S. Achtert. MLA Handbook for Writers of Research Papers. Second Edition. New York, New York: The Modern Language Association of America, 1984.

One of the classic texts on production of research papers. This guide covers all aspects of report production from how to begin researching a topic, to making note cards, outlining, and drafting the final paper.

Helix, Double. DNA. Portland, Organelle: Cellular Publications, 1980.

A double-stranded commentary on the molecule of life. This short text covers all the dimensions of DNA and its functions. Included is a discussion of replication and its importance to cell division.

Turbot and Bass. A Whale of a Fish Story. Flycast, Montana: Angler Publications, 1979.

A story that is hard to categorize as either fact or fiction. Written in first-person, the story has a ring of accuracy. However, some of the sizes of the fish supposedly caught in lakes and streams in Phoenix, Arizona, make the complete text hard to believe.

(continued)

Scribe's Name ______________________________ Date ____________________

Sample Bibliography *(continued)*

Newspaper Articles, Pamphlets, Interviews

"DEWEY BEATS TRUMAN." The Chicago Daily Tribune. Nov. 6, 1948.

A classic example of an "oops" in journalism. The article declared Thomas Dewey the winner of the 1948 presidential election. Final voting results, however, showed that Harry S. Truman had been elected.

Fungus, Amongus. Slime Mold as Pets. Mushroom Mycelium, Maine: The Penicillin Press, 1984.

This article is a do-it-yourself guide to finding, feeding, and propagating these yellowish organisms in and around the home. Specific dos and don'ts are spelled out. Types of materials to keep your mold away from are clearly defined.

Leo, Felis, Dr., Curator of Cats, San Diego Wild Animal Park. Personal interview, March 15, 1988.

This interview was a follow-up on the information from Lions in the Sahara Desert. *Dr. Felis answered questions on pride behavior and feeding habits. The interview, conducted just outside the lion enclosure, included a tour of that cavelike structure.*

Referencing

(Check with your teacher to see if this is part of your assignment.)

Referencing information is critical in a scientific report. Direct quotations, paraphrased statements, ideas taken directly from an author, lists, numbers or figures, and any information that is not common knowledge must be referenced. The referencing format to be used in this paper is one that is fairly common in college papers. Immediately after the information to be referenced, enclose in parentheses the following information: author's last name; the date of publication; the page number where the information is located. Study the form of the following sample: Claws. 1988:1225. If you have done enough research and have used the referencing format properly, each page of your report should have three to five references. Also, all of the entries in the bibliography should be referenced at least once in the text. (Fungus. 1988:6) The use of all bibliography entries will be checked as part of the grading standards.

General Science: Analysis

3. Energy: The Choice Is Yours

Teacher Instructions

We recommend using the information that follows as an introduction to this exercise. As the students attempt to solve the "Catch-22" problem of the drink cups, they will be easy to lead into the listing and ranking of the energy sources.

"As members of what has become a global society, we make choices every day in a wide variety of life areas. Some of our choices appear isolated but really have far-reaching consequences.

"For example, write down an answer to this question: *Which is worse for the environment, using paper cups or Styrofoam cups when you go on a picnic?*

"From a **landfill** point of view, it is hard to define which is worse. It takes many years for a Styrofoam (polystyrene plastic) cup to break down to any noticeable degree. Garbologists, people who study landfills, have shown that some types of paper can last for over forty years. When Bill Clinton and Al Gore were running for president and vice president, a newspaper was found in a landfill in Arizona with an article describing a Gore running for vice president. However, the Gore was Al Gore's father. And the date on the newspaper was 1952!

"From an **energy** standpoint, using the paper cup may be worse. It takes more energy to make a paper cup than it does to make a cup from Styrofoam.

"In some parts of the country the amount of **water** it takes to produce the two cups is a major environmental concern. Styrofoam cups require fewer gallons of water for production than paper cups. You may know from visiting a restaurant that practices water conservation that it takes three glasses of water to serve one glass to a customer: the first to serve, the second to wash, the third to rinse.

"So is reusing a glass or plastic cup better than using either paper or Styrofoam?"

The seven energy choices around which the exercise was designed are: **solar**, **fossil**, **nuclear**, **hydroelectric**, **geothermal**, **wave (tidal)**, and **wind**.

This is an excellent exercise to do as a values clarification exercise after the ranking of the energy sources. Students can follow the procedures outlined in the front of this book, use the consensus answer sheet, and move around the room as they line up under the number that matches their choice of energy for each of the four areas: **expense**, **pollution**, **practicality**, and **overall**.

Answers to questions:

1. Answers will vary. Check the students' charts to see if they were correct. There is always the possibility of a tie in the rankings.
2. Answers will vary. The *best* answer would be the energy source with the lowest average because it would have been rated at the *least offensive* overall.
3. Answers will vary. The important part of this exercise is the defense of the choice.
4. Answers will vary. The important part of this exercise is the defense of the choice.
5. Answers will vary. Some changes would be in the types of automobiles, the time it takes to go over long distances while traveling, and changes in wake up and bedtimes.
6. Answers will vary. Any of the energy choices would be possible to select. Probably nuclear will be the most common. Reasons for choosing to ban nuclear include waste disposal and contamination problems.
7. Answers will vary. Something to do with problems still to be solved should be addressed.

General Science: Creative Writing

4. Don't Get Bogged Down on This

Teacher Instructions

This assignment could be used while studying any one of the research areas, or as a cumulative assignment. You might also be able to cooperate with a world history or geography teacher on this and have the assignment be a part of two classes' work.

You do not have to use the time frame we supply in the story. It is just as acceptable to assign any age to the body. The period of world history your students are studying when you give this assignment, whatever time period that is, might be the most appropriate time period to use.

There are some excellent resources available on frozen or naturally mummified people. One of the books, *Bog Man* by Don Brothwell, has a wealth of information on the bog people throughout Europe. In the February 1993 issue of *Popular Science* magazine, there is an article on the frozen body found on the Swiss-Italian border. The PBS series *Nova* did a program on the same body, "Iceman," in November/December 1992. *Nova* did an excellent job of explaining the era in which the iceman lived.

Grading of this assignment is based primarily on the completeness of the letter in regard to the assigned topic. The value for the students is in the comparative research that they do.

An alternate way to group students after the assignment has been written is to have all of the students who did the same topic group together. Each of those student groups could have a discussion after listening to all of the group papers. Consensus could be reached on a value ranking of a combined list of changes presented by each writer.

General Science: Critical Writing

5. Science in (and as) History

Teacher Instructions

This assignment can be done at any time of the year. Your school librarian will probably be happy to round up or make a list of all the books at the school dealing with this type of topic and have them available for your students. If you have lots of students, you might consider rotating this assignment through the semester, class by class. This will allow better student access to the available resources, and might maintain your sanity, since you would not be reading all the papers at one time.

This is a good assignment to team teach with a social studies teacher. Students could select scientists from the period in history that they are currently studying. This would reinforce the cross-disciplinary nature of the assignment. You could also consider teaming with an English teacher. They usually have report formats that the students could use to practice their English skills in your class. Sometimes English teachers provide sociological background for authors they are studying, as well. Learning about a seventeenth-century French author and writing about a seventeenth-century French scientist might be a good combination.

You can allow students free reign on selecting their scientist. This has the potential of being good or bad. It is good because students cannot complain that the scientist you assigned them was so boring. It is bad because many students could have a hard time deciding on a scientist or lots of students might do the same person. (How many reports on Pasteur do you want to read?)

We recommend making a list of scientists from which students can choose. Allowing a maximum number of students per scientist will help insure that you do not get fifty percent of the students working on the same scientist. A few minutes of looking through the references that the librarian collects for the students should provide you with enough information to construct such a listing.

General Science: Creative Writing

6. My Life as a Fish

Teacher Instructions

This assignment is specifically designed for portfolio use. After the assignment in pencil has been peer-graded, it should be recorded and returned to students for final revision. The final draft should then be placed in the student's portfolio along with the parts of the assignment turned in previously.

You must check that each of the students has completed all three parts of the assignment. If you use a rubber stamp on the paper to show completion of a part, then the presence of three stamp marks is all that is needed to give a grade for completeness at the end of the assignment.

Students will need access to a biology, life science, or vertebrate zoology book to get the information necessary to answer the original questions. An alternate resource would be an encyclopedia.

Answers to Questions

1. Fish use their fins for swimming. (a) The caudal (tail) fin is most important. (b) The caudal fin moves side-to-side in fish.
2. The gills (or gill filaments) of a fish are used to extract oxygen from the water in which the fish is swimming.
3. Mucus reduces friction for fish movement; it provides lubrication for the scales as they rub against each other during movement; and it is antiseptic, providing protection against fungal and other infections.

When checking the student papers for following instructions, be sure that part 2 includes complete sentences. Also, be sure to check for the four points (a–d) in part 3.

Instructions for student evaluation of student stories are given in "Life as a Bird Embryo," an exercise in *Cranial Creations in Life Science.*

Meteorological Science: Analysis

7. Rain, Rain, Go Away!

Teacher Instructions

The data in this exercise is actual data from the National Weather Service.

This exercise is designed to allow students to apply a variety of data analysis techniques to a large amount of data. The two types of graphing required could be supplemented by a line graph of one or more decades, if desired. If you choose to do the line graph, we recommend having students label the *x*-axis as shown here:

x --|-- --|-- --|-- --|-- --|-- --|-- --|-- --|-- --|-- --|-- --|--

1900 1901 1902 1903 1904 1905 1906 1907 1908 1909 1910

1950 1951 1952 1953 1954 1955 1956 1957 1958 1959 1960

1980 1981 1982 1983 1984 1985 1986 1987 1988 1989 1990

This type of *x*-axis will allow the students to graph thirty years of data in an economy of space on the graph paper. Using different colors for each year is a good way to show comparison, and this format is an alternative to the bar graph students do in question 9.

You will need to have your city's annual rainfall total from last year for the students to use in question 10.

Rainfall data given for the various biomes are approximations and are not intended to be taken as the only definitive rainfall totals for those areas.

Answers to Questions

1. Answers will vary. Students should give some reason for the amount estimated. Most students will probably estimate a low amount.
2. Answers will vary. This will depend on the sophistication of your students. **Chaparral**, **Mediterranean**, or **desert** are all used by various books.
3. Answers will vary. The answer should be the amount of rainfall that fell in their birth year. You might have the students put their birth year by the amount so you can check.
4. A. a. 0.00–3.00 inches (0 years)
      b. 3.01–6.00 (17 years)
      c. 6.01–9.00 (28 years)
      d. 9.01–12.00 (33 years)
      e. 12.01–15.00 (11 years)
      f. 15.01+ (11 years)
   B. Range d., 33 years
   C. This should be a histogram.
5. 1940–41 had the highest amount, 24.74 inches.
6. Average = 9.93 inches (25.22 cm)
7. From the data given, **Desert**.
8. 100 year total = 992.87 inches which would be 4.96 years in a TRF at 200 inches per year.

9. Students should provide a bar graph. The 1950's had two years of heaviest rainfall of the three decades plotted.
10. 1939–1940 to 1943–1944 is the five-year period with the heaviest rainfall. The total for that five-year period was 74.66 inches. The five-year average was 14.93 inches, 5 inches above the 100-year average of 9.93 inches.
11. 1958–1959 to 1962–1963 is the five-year period with the least rainfall. The total for that five-year period was 29.80 inches. The five-year average was 5.96 inches, 3.97 inches below the hundred-year average of 9.93 inches.
12. Answers will vary. Unless you live in a desert-like area yourself, you will probably have more, and maybe *much* more rain than San Diego's annual average.

Meteorological Science: Laboratory Skills

8. Fluffy as a Cloud

Teacher Instructions

DAY 1

Each team of students will need: one thermometer, one shiny can (no label), two to three large ice cubes, water, paper toweling.

Answers to Questions

1. Answers will vary.
2. Answers will vary. This temperature should be only a few degrees higher than the temperature in question 1.
3. This answer should be the average of the answers to questions 1 and 2. Check to see that all answers are in either degrees F or degrees C.
4. The relative humidity is 100%.
5. Saturated
6. Less than 100%
7. Answers will vary. Rain would usually form, although snow or hail could form if the dew point is below freezing.

DAY 2

Be sure that there is some type of cloud formation overhead on the day you do this portion of the activity.

There is a very real chance that the calculated altitude in question 6 will not match the type of cloud formation overhead. Do not panic. The questions allow this.

In addition to the materials from Day 1, students will need reference pictures of cloud types and information on the altitudes at which different cloud types are found. Many earth science books have these kinds of pictures and data.

Answers to Questions

1. Answers will vary. The three main types of clouds are **cumulus**, **stratus**, and **cirrus**.
2. Answers will vary. Check the reference materials your students are using for more specific figures. As a general guideline, cumulus clouds are found at an altitude of 2.4 to 13.5 kilometers, stratus clouds are found at an altitude of 2.5 km, and cirrus clouds at 6.0 to 12.0 km.
3. Answers will vary. All teams doing this activity at the same time should have nearly the same figure for this answer.
4. Answers will vary. All teams should have nearly the same figure for this answer. You might want to spot check answers to make sure every team has done its calculations correctly.
5. The greater the altitude, the lower the air temperature.
6. Answers will vary. All teams should have much the same answer.
7. Answers will vary. All teams should have much the same answer.
8. Answers will vary. Students should use the word "agree" or "disagree" in their answer and explain that the observed clouds were the same type or a different type from the clouds that would form at their calculated altitude.

Meteorological Science: Creative Writing

9. My Life as a Cloud

Teacher Instructions

This assignment allows the students some creative license as they report the sequence of steps in cloud formation. Encourage students to write in the first person and give some of their feelings as they change altitude and condense.

The sequence of events for this assignment would be:

1. Assign the story.
2. Allow time for research and writing (variable, depending on your curriculum and depth of coverage expected by the students).
3. On the day the story is due, collect it from the students and check each one for completeness.

4. Return individual papers to students and assign groups of three to five students for the group portion of the activity.

Information for the writing of this story can be found in any earth science textbook or meteorology book. Succinctly, the steps in cloud formation are:

1. As warm air rises, it cools down until it reaches its saturation point.
2. At the saturation point, water droplets start to condense around a condensation nucleus.
3. Water droplets combine to form larger droplets and a cloud is formed.

Answers to Questions

1. Water droplets join together to form a cloud when rising air cools off enough to reach its saturation point. (Cool air can hold less water than warm air, so cool air more quickly reaches its saturation point.)
2. Water droplets join together to form larger droplets in the process of condensation. They condense around a particle of dust or smoke, etc., known as a condensation nucleus.
3. The three basic kinds of cloud formation are (1) **cumulus**—fluffy, white clouds with flat bottoms; (2) **stratus**—smooth, gray, layered clouds; (3) **cirrus**—feathery, wispy clouds.
4. Cumulus clouds are generally found at an altitude of 2.4 to 13.5 km; stratus clouds at 2.5 km; and cirrus clouds form between 6 and 12 km.
5. Answers may vary. Clouds might change type if air is pushed higher or lower in altitude by advancing cold or warm fronts. For the same reason, clouds might also evaporate and become smaller, or become larger if more water vapor condenses.
6. Fog

Earth Science: Creative Writing

10. Let's Get Mineralized

Teacher Instructions

This assignment allows the students some creative license as they report the sequence of a geologic process. Encourage students to write in the first person and give some of their feelings as they fossilize and when they are discovered.

The sequence of events for this assignment would be:

1. Assign the story.

2. Allow time for research and writing (variable, depending on your curriculum and depth of coverage expected from the students).
3. On the day the story is due, collect it from the students and check each one while the students are answering the questions individually.
4. Assign groups of three to five students for the group portion of the activity.

Information for the writing of the story part of this assignment can be found in any earth science textbook or a recent book on dinosaurs or fossils. Briefly, the steps of fossil formation are . . .

1. A submerged body is covered with **sediment**.
2. Pressure from above turns the sediment into rock.
3. Percolating minerals diffuse through the bones and are deposited in the skeleton. The bones themselves become rock.
4. The **fossil** is exposed to all of the geologic forces, as is the rock in which it is found.

An excellent way to integrate biology (genetics, genetic engineering) with this earth science material would be to discuss an approach like that of Michael Crichton in his best-seller *Jurassic Park.* In the book, dinosaurs are reconstructed with genetic engineering techniques using DNA samples from fossils.

Answers to Questions

a. Covering the body (bone, carcass, animal) with sediment.
b. The oldest fossils (bacteria) are 3.5 billion years old. Fossil mammoths found frozen in ice are from 8,000–10,000 years old. Several million years would be a reasonable time estimate for a mineral-replacement fossil to form.
c. Bones and teeth are the best at fossilizing. Imprints of skin and feet can also fossilize.
d. Most probably the body would have been eaten by scavengers. Decomposition would have destroyed the rest of the body, leaving no trace of the existence of the allosaurus, except as fertilizer for the soil. But, the dinosaur could have become part of a fossil fuel deposit as well.
e. Both allosaurus and camptosaurus have been uncovered in the western United States. They were active in the Jurassic period.
f. There have been many recent finds of dinosaur nesting sites complete with fossil eggs in various stages of development. The shell of the egg and the bones of the embryo would be suitable material for fossilization to occur.

Earth Science: Analysis

11. My Fossil's Older Than Your Fossil

Teacher Instructions

In the first part of this exercise, the students' only clues are the external characteristics of the organisms. This part of the challenge should prompt some excellent discussion about what relationships exist between the organisms.

It is very important to emphasize that there are three branches from the original *tortugis.* It is equally important to tell the students that there are more than three branches on the whole family tree.

The fact that an organism has been found as a fossil does not mean that it cannot still exist today. And the fact that there are no fossils between two organisms does not mean that an intermediate form could not have existed.

The family tree included in these notes is the way that *we* invented the creatures. However, it is *not* the only way that the organisms could be arranged in a logical fashion. In grading student work, you should be looking for reasonable explanations.

When students receive the diagram of the geologic strata and the depth of each fossil located (Part 2), be sure that you encourage them to work on the questions before they spend too much time revising their family trees.

One way to treat this exercise is to have groups of students do the original family tree, the revised family tree, and the answers to all of the questions. An alternative would be to have the original group of students do the original family tree, the revised family tree, and the answers to questions 1–4. Each student could then be assigned question 5 as an individual assignment. If you choose the second option, it is imperative that you allow students some time to sketch the two organisms between which they are going to invent and place their missing link. You might even allow students to take home the cutout pictures of the organisms.

Answers to Questions

1. Answers will vary depending on student's version of tree. The names here should match those of the organisms closest to *T ancientis.*
2. Answers will vary. The names here should match those on the *highest* ends of the tree limbs.
3. Yes. Not all organisms die in conditions that produce fossils.
4. Answers will vary. The reasoning of the answers is the most important quality to look for here.

5. Answers will vary. The answer to (a) should include a picture. The name chosen in (b) should represent some key feature of the missing-link organism designed.

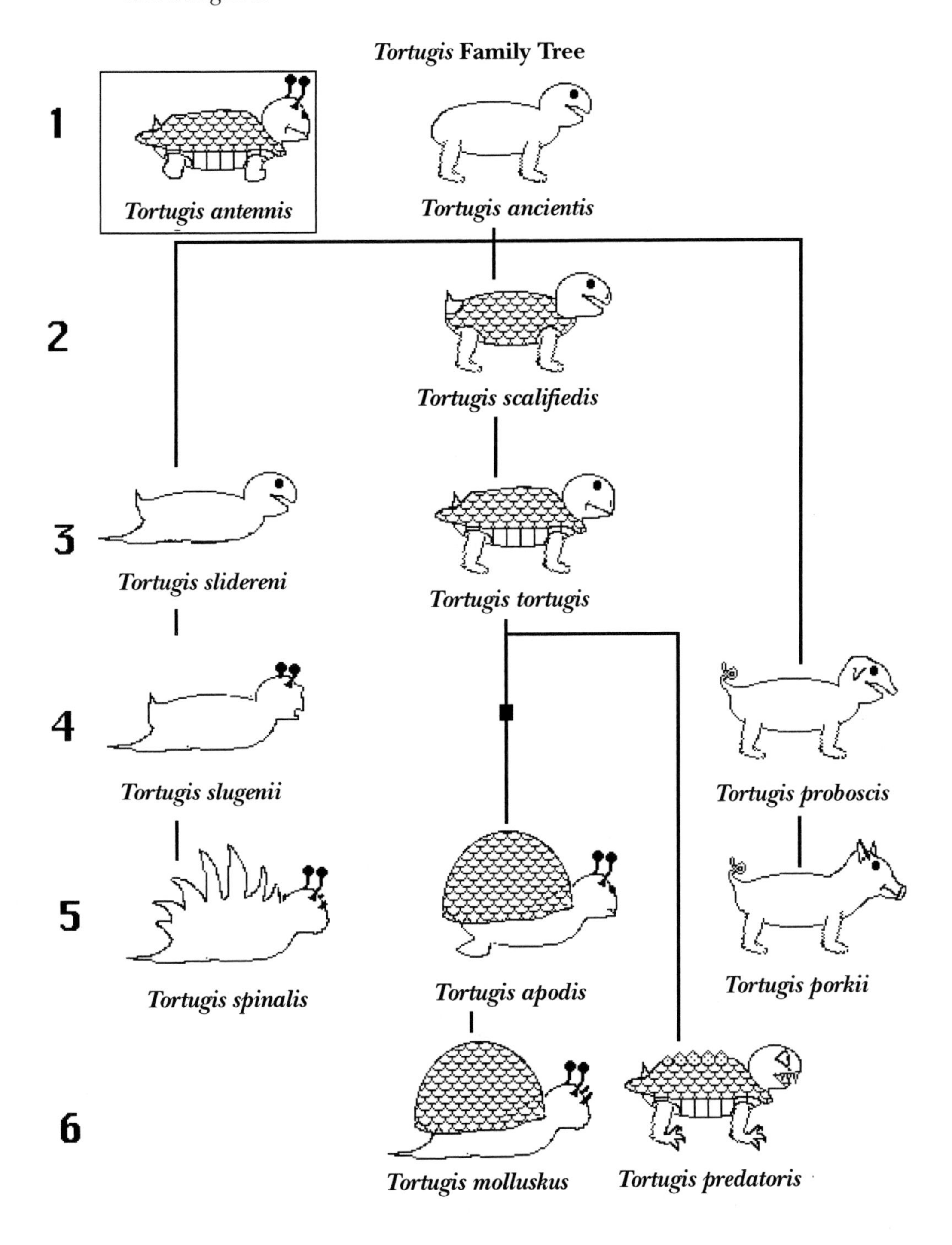

Earth Science: Creative Writing

12. There'll Be a Hot Time in the Old Town Tonight!

Teacher Instructions

This assignment could be used in a variety of ways. Students should show their understanding of the volcanic process by determining what would and would not survive the intense heat of the lava flow.

National Geographic has published at least two articles on discoveries at Pompeii and Herculaneum. By injecting latex into the cavities left behind in the hardened ash after bodies had decayed, archeologists have been able to recreate the final moments of a variety of people in these cities. It is recommended that you have these issues of the magazine or other source material on hand for the duration of this assignment.

Answers to Questions

1. Anything made of hard material like rock or shell would survive under the protective layer of ash from the eruption. Lava is, of course, hot enough to melt even rock, so anything sticking out of the ash layer—grass from the huts, trees, or rock material—would have been incinerated by the lava flow.
2. Protection by the layer of ash.
3. Flesh, wood, leather, coconuts. They would be burned by the lava.
4. The diagram can show any layout of the village as long as it has about twelve circular outlines of huts (still discernible by the charred posts), one fairly complete rectangular foundation of coral stones, and a variety of body outlines, shell jewelry, and stone or metal implements, without the wooden handles.

Earth Science: Art

13. "Time" for a Vacation Trip

Teacher Instructions

This could be a take-home assignment or an individual or cooperative assignment in class. If you choose to do this as a cooperative, in-class assignment, the following guidelines are recommended.

- Have students work in pairs or trios.

- Allow students only a portion of the period for completion of this task. Twenty-five minutes should be adequate.
- Establish accountability by requiring the entire group to show its poster to the class.

If this assignment is done in class, you will need to provide colored pencils or pens and paper.

We recommend limiting the number of students who choose any single geologic period in each class. You might also want to limit which of the periods you want the students to choose from and have them sign up for the period they wish to do their poster on.

Grading this assignment is fairly subjective. We give 100% to outstanding posters. We award an A grade to any poster that was produced following the instructions completely. Major deductions should be made for:

- no color
- incorrect number of slogans
- incorrect style of slogans
- destination not given or in wrong location
- wrong size

You can use this as a way to give extra credit if you are so inclined. Simply lower the possible value of the assignment by 40%. Award the revised full credit amount to anyone who follows the instructions. Award up to 40% extra credit for the best, most creative posters you get.

It is very easy to add a short written portion to this exercise by assigning a one-paragraph description of the geologic period to be attached to the back of the poster.

Earth Science: Analysis

14. Sounds Pretty Deep to Me

Teacher Instructions

This exercise is designed to simulate what computers would do with sounding data collected during an ocean-floor mapping journey.

We recommend dividing the class into pairs for this exercise.

Each pair of students gets one of the sets of coordinates A–R. They plot the points on a piece of graph paper and connect the dots to form a bottom profile. After plotting the points and drawing their profile, they compare their results with the original map of the ocean floor that you put together (see below). They try to determine which section of the ocean floor they reproduced. Some of the sections are almost exactly reproduced by the coordinates. Other sections, especially the sections with 90-degree cliffs and/or overhangs, do not reproduce extremely accurately by using the coordinates, but that is true in actual sounding as well.

The original map of the ocean floor developed for this exercise has the ocean bottom sections in the following order: 1=E, 2=Q, 3=O, 4=M, 5=B, 6=I, 7=K, 8=L, 9=F, 10=H, 11=C, 12=P, 13=G, 14=A, 15=J, 16=D, 17=N, 18=R. The profiles that follow are printed, two to a page, in this order. We recommend photocopying each of the pages, and arranging them in numerical order somewhere in your room. After the students complete their profile they can match their plots with the original numbered sections to check for accuracy.

You do not have to use all of the profiles. If you do not use them all, you might not want to put all of the profiles up as part of your ocean floor. You could post only those sections that the student groups have to work on. But posting the entire ocean floor and using only some of the sections with your students will provide a greater challenge than posting only those sections with which students are working.

Copy the general instruction pages for each student or group. Then copy those sets of coordinates you wish to use. The coordinates are arranged in two sets per pair in the student section to conserve space.

Depending on a number of factors, including your own preferences, the personality of your class, and the availability of colored pencils, you may or may not wish to have the students complete question 8. Students are asked to check with the teacher before coloring in their section of the sea floor. It could be fun.

Answers to Questions

1. Answers will vary. This depends on which set of coordinates the student group was assigned.
2. Answers will vary. Reasons could include: not enough data points, inaccurate measurements in the ocean, inaccurate measurements on the graph paper.
3. Answers will vary. Adding more data coordinate points to plot would be the most efficient method.
4. Answers will vary. Changing to sonar would give the most accurate picture. Doing more sounding points would also help improve accuracy.
5. a. There will be a more accurate picture of that section of the ocean floor.

   b. Many features (hills, valleys) would be missed in the profile.
6. Answers will vary. This is a number. Check the profile to see that it matches the minimum depth coordinates. Remember, the scale of the profile is 1" = 100m.

7. Answers will vary. This is a number. Check the profile to see that it matches the maximum depth coordinates.
8. This is an optional portion of the assignment that will allow the students to have a little fun. After all of the individual profiles are colored, they could be placed on a bulletin board to form a picture of the "entire ocean." This could be humorous, since each group will have its own idea of what the ocean floor looks like. You might have a tropical sea floor next to a portion of the Arctic ocean bottom.

Earth Science Analysis

14. Sounds Pretty Deep to Me

Original Ocean-Floor Profiles

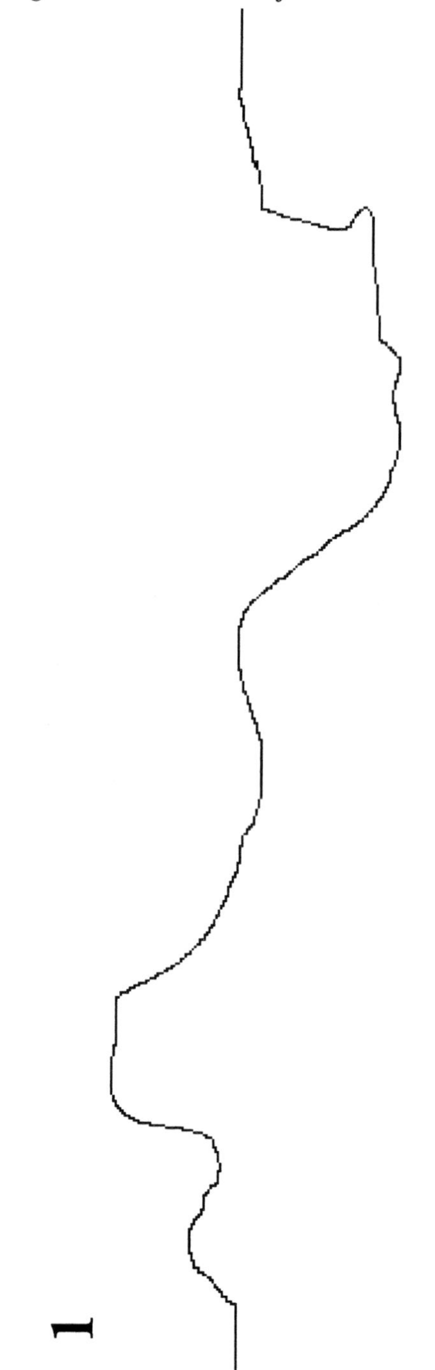

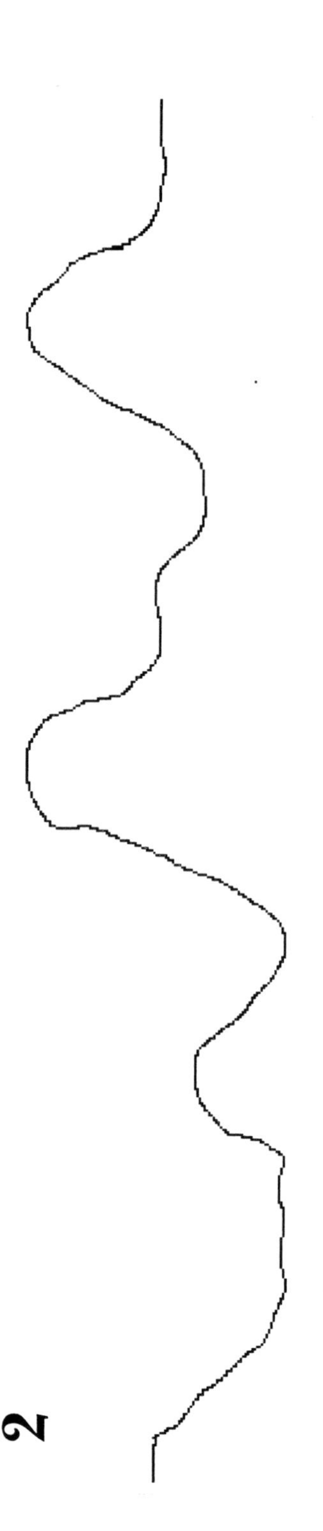

Original Ocean-Floor Profiles

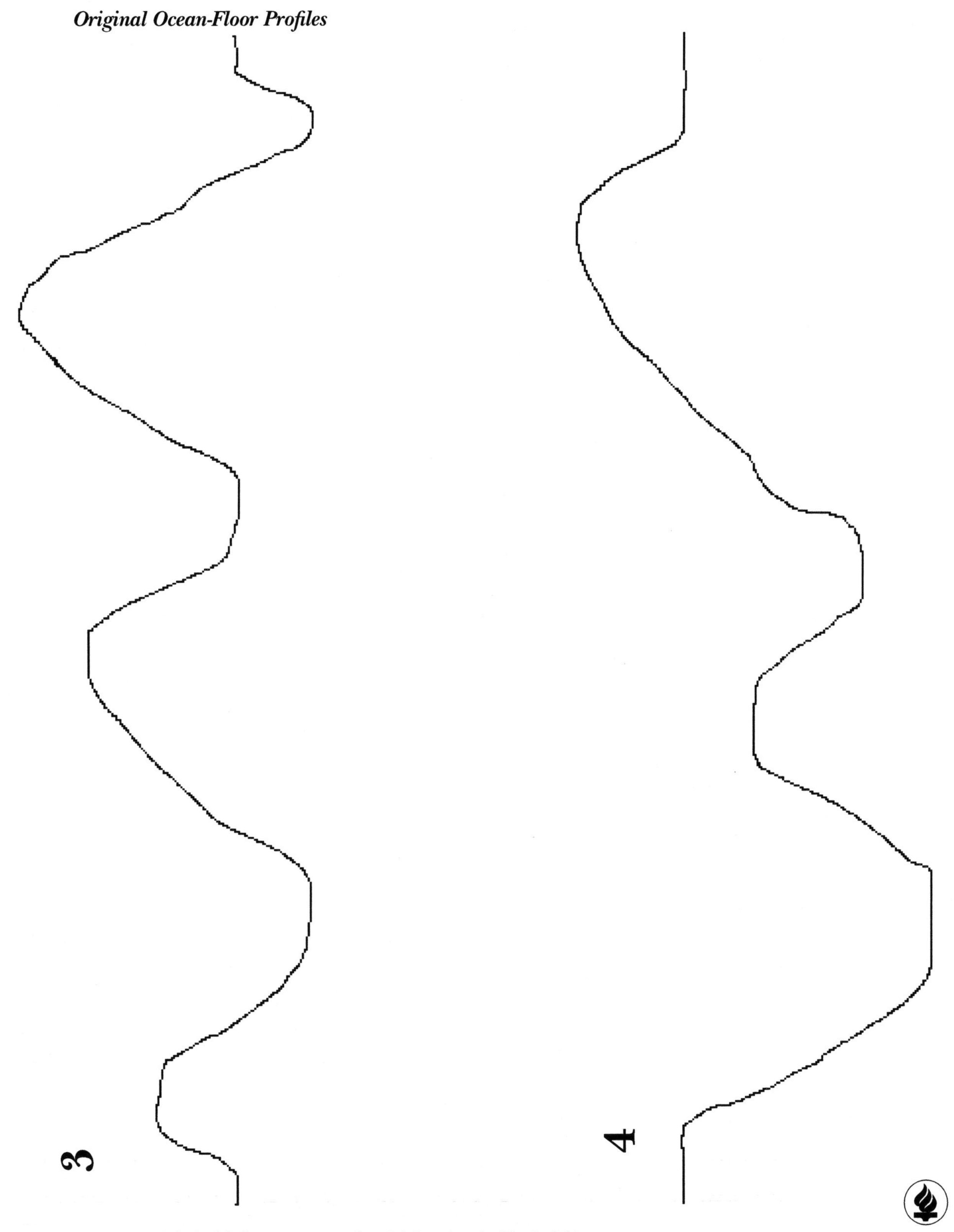

Original Ocean-Floor Profiles

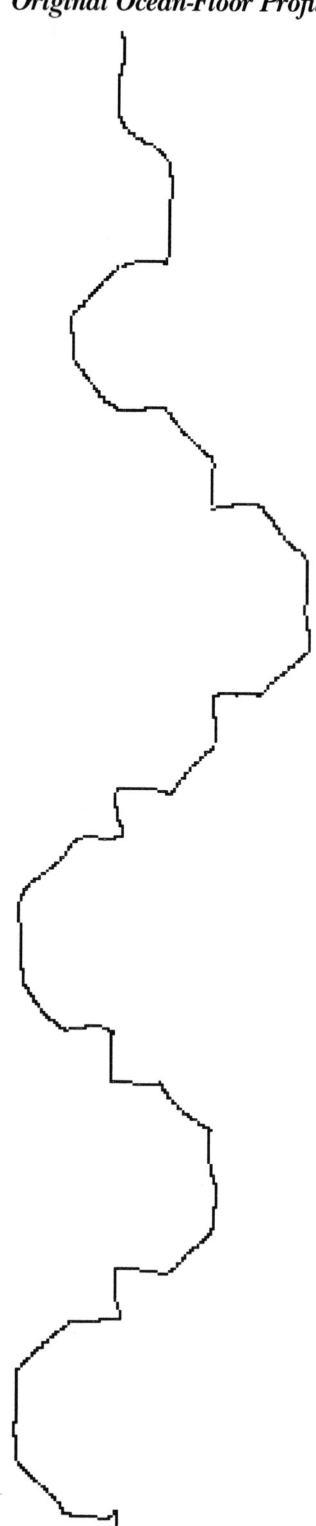

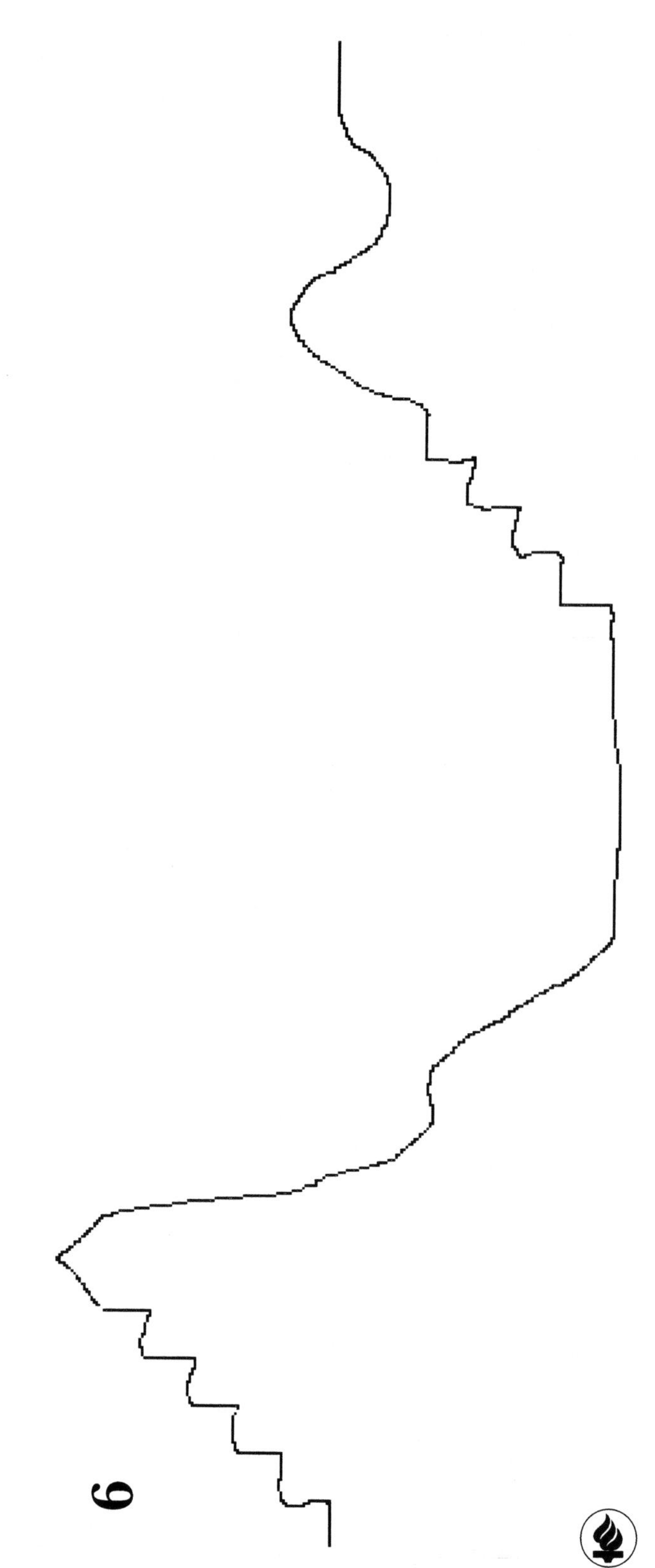

Original Ocean-Floor Profiles

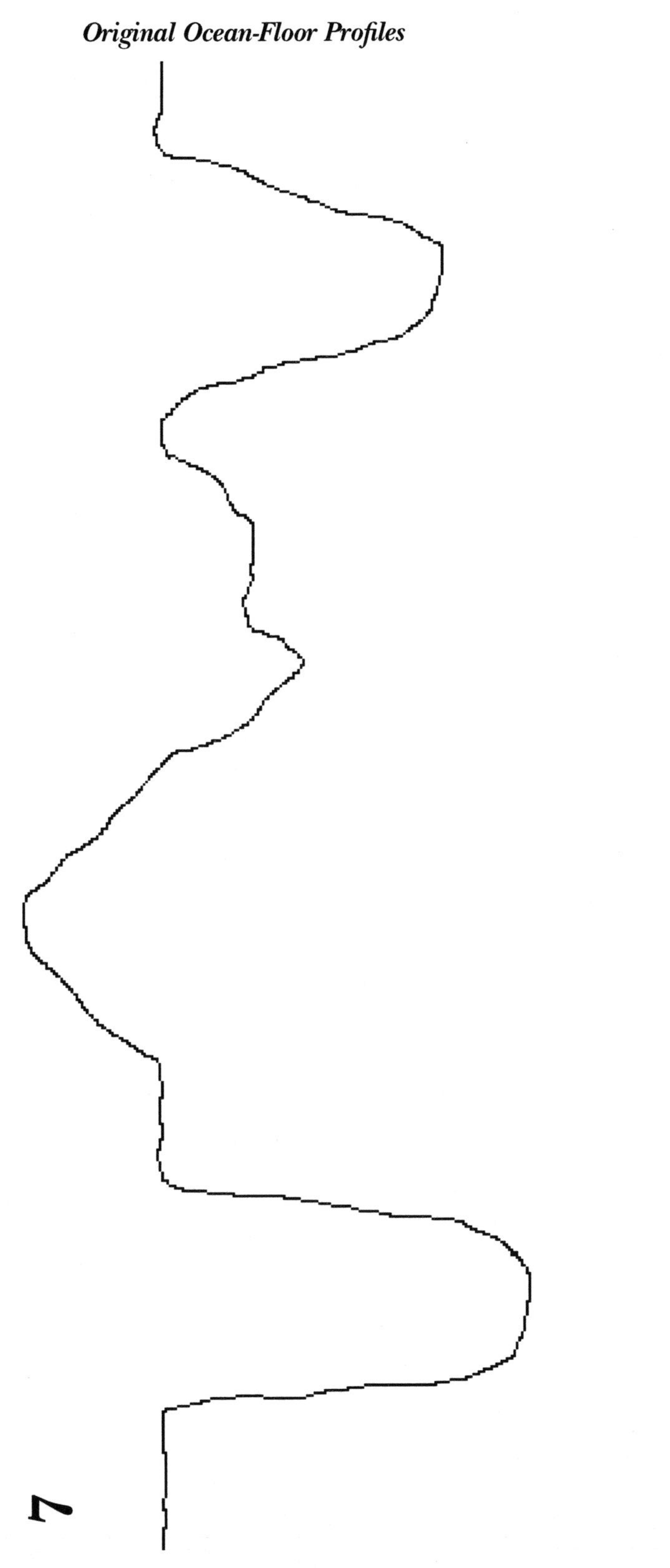

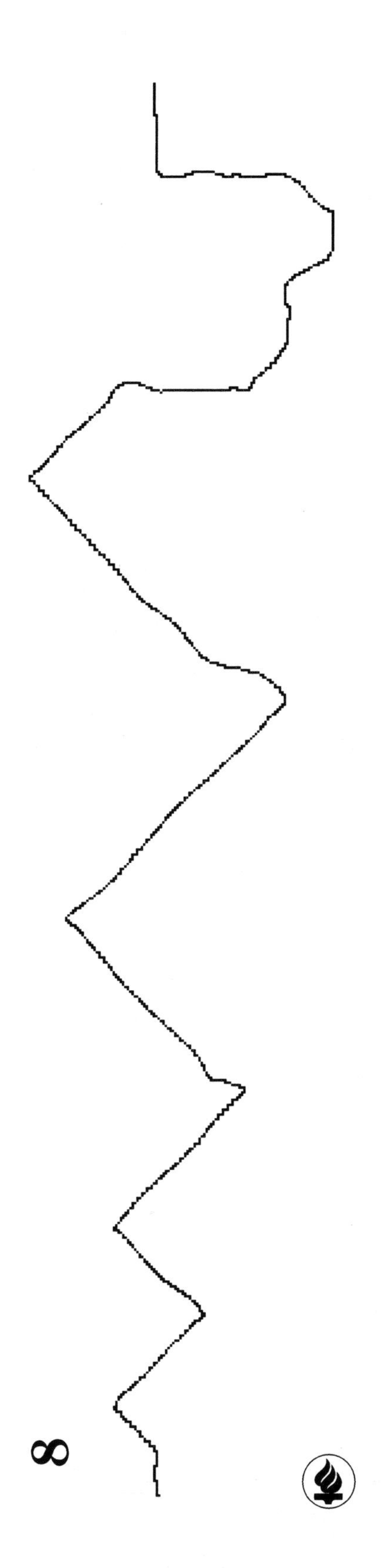

Original Ocean-Floor Profiles

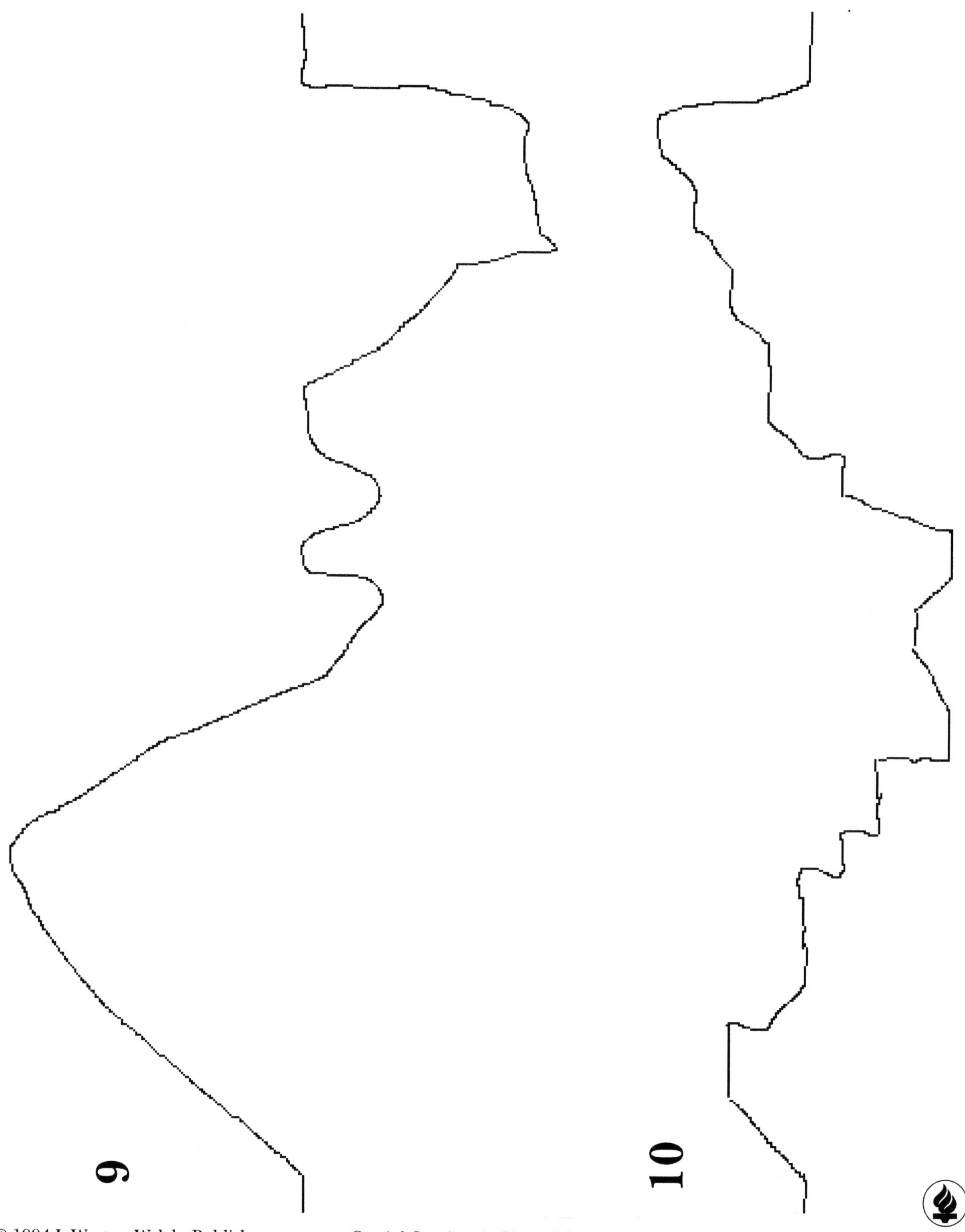

Original Ocean-Floor Profiles

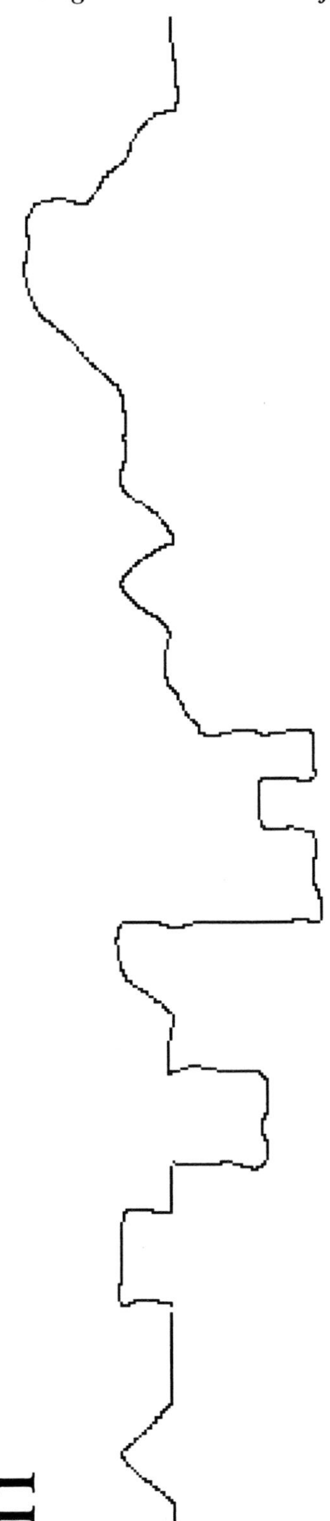

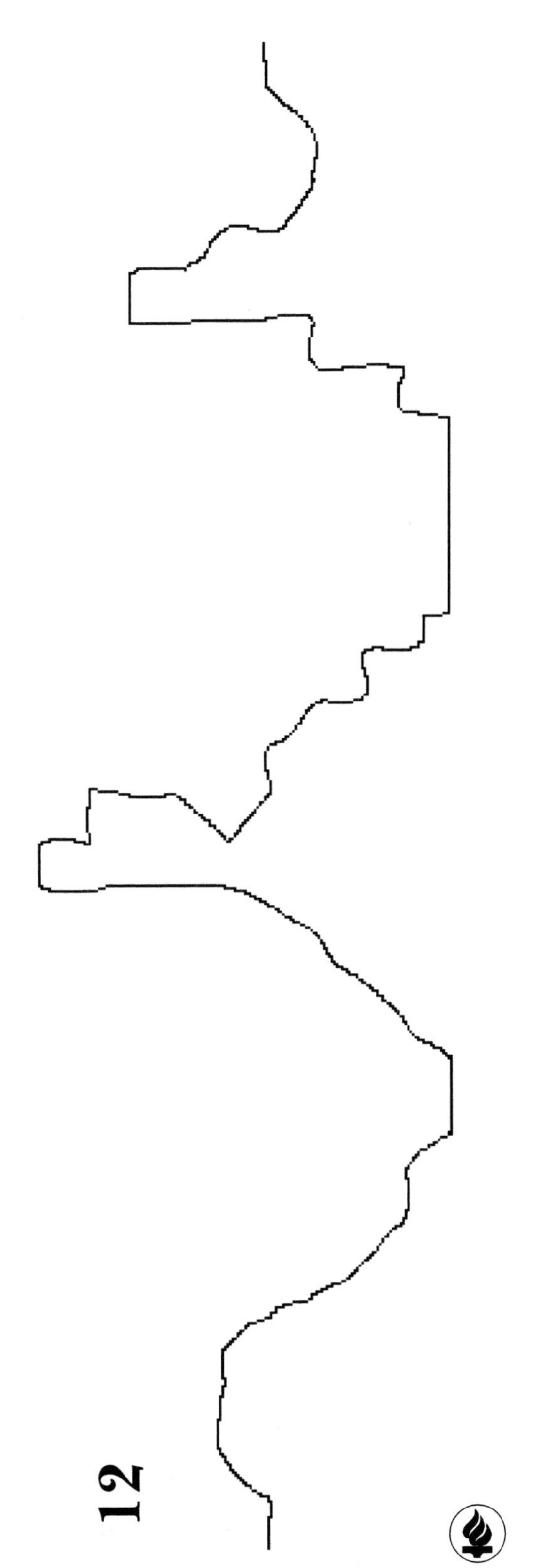

Original Ocean-Floor Profiles

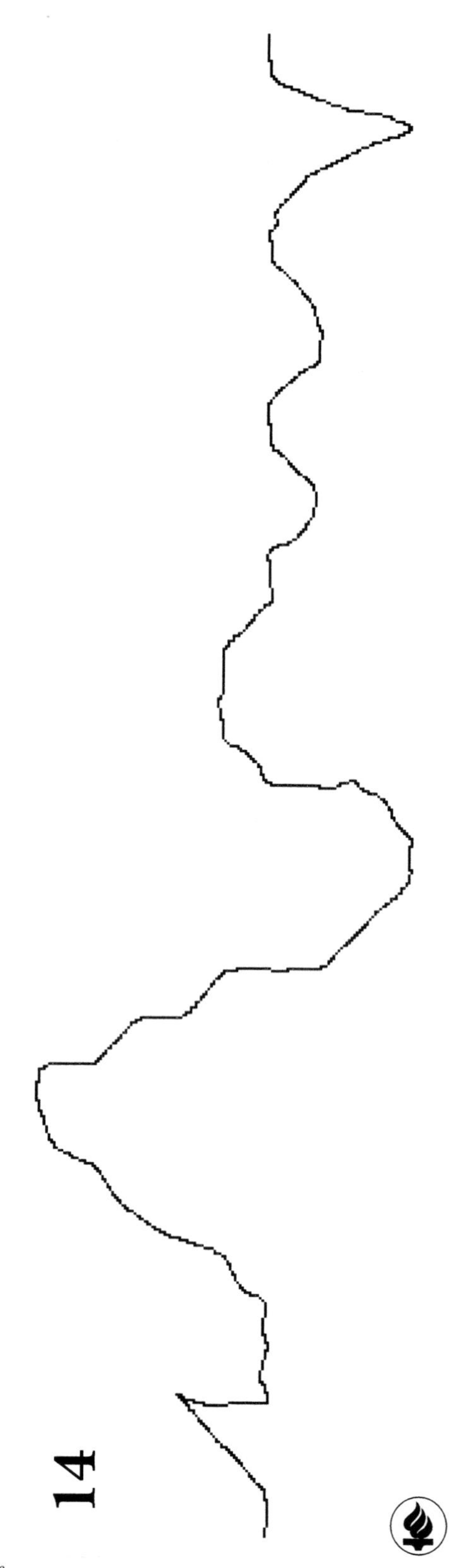

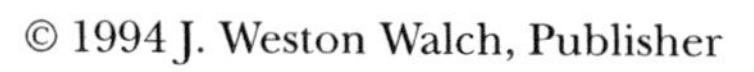

Original Ocean-Floor Profiles

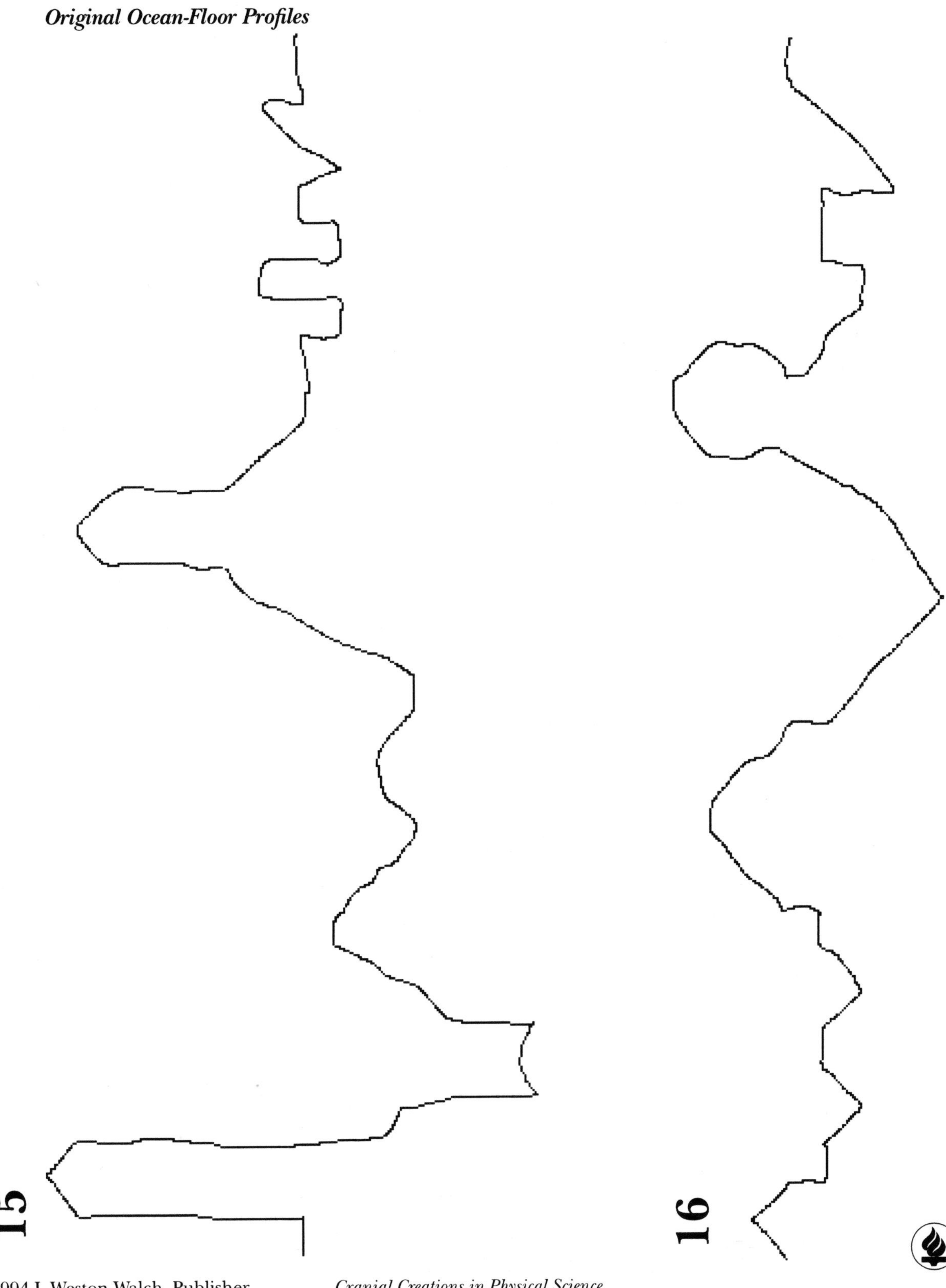

Original Ocean-Floor Profiles

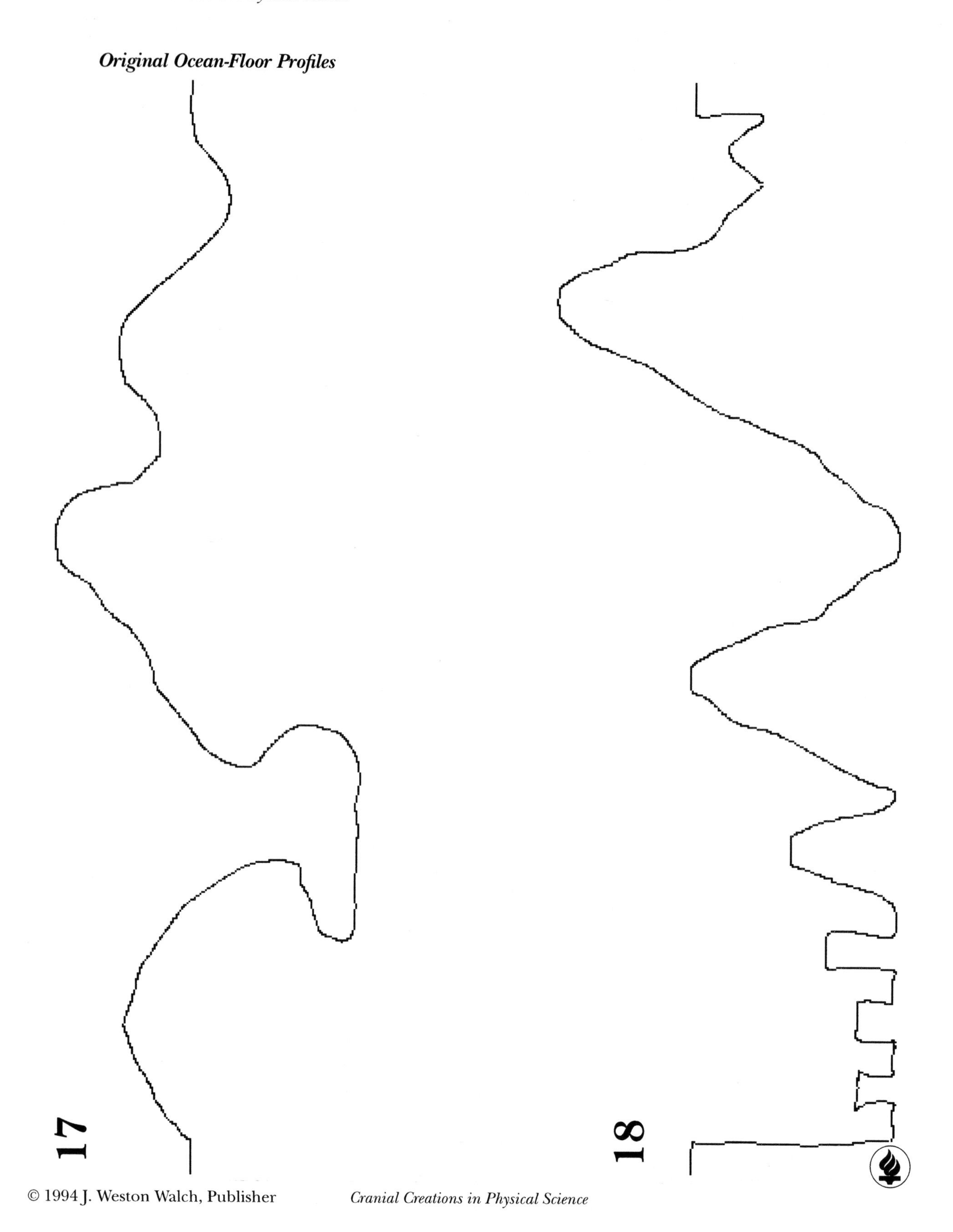

Earth Science: Art

15. Move Over, Gary Larson

Teacher Instructions

You should have colored pencils or felt pens and unlined paper ready for students to use in this exercise.

Grading of this Cranial Creation is subjective. Basing the grade on appropriateness seldom results in any complaints.

A good indicator of the appropriateness of the cartoon is the class reaction to the opaque projection of the cartoon onto the screen. If the students in the class do not understand the cartoon, but you do (or vice versa), it is probably still a reasonable cartoon. If neither you nor the students understand the cartoon, well . . . .

You will probably find the political cartoons the vaguest.

This is a good way to do a unit review. Students have to demonstrate factual knowledge and infer appropriate relationships.

To help focus your students, you could put a list of acceptable topics on the board. Student groups could sign up for a topic before they begin their cartoons. This would allow you to regulate the number of cartoons on any one topic. It would also allow you to make sure that the topic areas you want to emphasize are not neglected, without being heavy-handed.

Don't be shy about using this exercise after a weather or space unit or other more traditional earth science unit. It provides an avenue of creative expression for students of all levels of ability.

It might also be interesting to have all of the students do their cartoons on a single topic, for instance earthquakes, to see the breadth and depth of insight and understanding in the entire class.

Earth Science: Creative Writing

16. Rock On!

Teacher Instructions

This assignment allows the students some creative license as they report the sequence of a geological process. Encourage students to write in the first person and give some of their feelings, both as they become the rock and when they are discovered.

There are several options for how to proceed with this assignment.

One of the possible sequence of events for this assignment would be:

1. Assign the story and form the groups of three students for the activity. Collect the notebook paper with names and types of rock they will become.
2. Allow time for research and writing (variable, depending on your curriculum and depth of coverage expected from the students).
3. On the day the story is due, collect it from the students and check each one while the students are answering the questions individually.
4. Return the cover sheet with names and rock type when the students begin to work on the group portion of the assignment.
5. Allow time for students to teach each other the information.
6. Give the quiz. One way to reduce the threat of the quiz is to make it a group quiz. If you do this, you might try allowing only written communication during the quiz portion of the assignment. This keeps the groups focused on *their* group members' ideas. You could be nice and allow the group time to work on the quiz after they have done their own work on the quiz as individuals. This could provide a form of reward while continuing the cooperative nature of this assignment.

You could leave out the individual assignment on the due date. You could leave out the quiz and treat this assignment like another creative writing story assignment and use the suggested procedure at the beginning of this book. You could think of other options, too.

Information for the writing of the story part of the assignment can be found in any earth science textbook or a book on rocks and minerals. The steps of rock formation are answered in the questions posed in the individual in-class assignment.

Answers to Questions

Igneous Rocks

a. Molten **magma** from inside the earth begins to cool.
b. Heat is the primary force. Intense heat melts rock forming magma. The magma cools, forming **igneous rock**. **Intrusive** igneous rock which cools inside the earth, forms slowly compared to **extrusive** igneous rock, which cools rapidly at the earth's surface.
c. Six minerals are common to igneous rock: **quartz**, **feldspar**, **pyroxene**, **amphibole**, **olivine**, and **mica**.
d. Answers will vary. Extrusive rock might have been deposited by a volcano. Either type of igneous rock could have been pushed to the surface by folding.

e. Answers will vary. Some place close to a volcano or where magma oozes from the crust. Hawaii would be an idyllic place to be found.

Sedimentary Rocks

a. Bits and pieces of eroded or broken rock, sand, mud, or broken shells of organism are deposited somewhere, usually under water. Other **sediment** stacks on top of this bottom layer.

b. **Pressure** is the primary force. Over hundreds of thousands to millions of years, tremendous pressures on the layers fuse the pieces together into sedimentary rock.

c. Answers will vary. **Sandstone** usually contains **quartz** (the primary component), **zircon**, **garnet**, **rutile**. **Shale** contains **clay** and **mica**. **Siltstone** contains **quartz** and **clay**. **Limestone** and **chalk** contain **calcium carbonate** (**calcite**). **Rock salt** contains **halite**. **Gypsum** contains **gypsum**.

d. Answers will vary. **Erosion** of higher layers probably exposed the rock. That layer got so "high" because the river eroded through other layers and is lower than it was. Folding might also have pushed the layer up.

e. Answers will vary. Some place that was once the bottom of an ocean. Grand Canyon comes to mind.

Metamorphic Rocks

a. Some existing type of rock is placed in a location, usually 12–16 km below the earth's surface, where heat, pressure, and chemical reactions can act on it.

b. **Pressure**, **heat**, and **chemical reactions** between components of the rock are the primary forces which act fairly equally or vary from type to type of metamorphic rock. In any case, it takes all three forces to form metamorphic rock.

c. Answers will vary. If metamorphic rock forms from (1) **sandstone**, it would contain **quartz** (the primary component), **zircon**, **garnet**, **rutile**; (2) **shale**, it would contain **clay** and **mica**; (3) **siltstone**, it would contain **quartz** and **clay**; (4) **limestone** and **chalk**, it would contain **calcium carbonate** (**calcite**); (5) **Igneous rock**, it would contain **quartz**, **feldspar**, **pyroxene**, **amphibole**, **olivine**, and **mica**.

d. Answers will vary. **Erosion** of higher layers probably exposed the rock. That layer got so high because the river eroded through other layers and is lower than it was. Folding might also have pushed the layer up.

e. Answers will vary. Any place where marble is found. Just outside a marble quarry in the north of Italy wouldn't be a bad location.

Answers to the Quiz Questions

1. Metamorphic
2. Sedimentary
3. Igneous
4. Sedimentary
5. Metamorphic
6. Igneous
7. Metamorphic
8. Igneous
9. Sedimentary
10. Sedimentary

Scribe's Name ______________________ Date ______________

Earth Science: Creative Writing

16. Rock On!

Quiz

1. What type of rock can start as any of the three types?

2. What type of rock is formed by pressure or cementing together?

3. A volcano is most likely to have produced which type of rock?

4. Which type of rock forms from layers of materials?

5. Which type of rock is always formed within the earth's crust?

6. Cooled magma is which type of rock?

7. Which type of rock takes three forces working together to form it?

8. Heat is the primary force in forming which type of rock?

9. Sandstone and limestone are examples of what type of rock?

10. Which type of rock can be made up of fragments of fossilized remains?

Earth Science: Art

17. Spend Some Time in the "O" Zone

Teacher Instructions

This is a good way to review any work you have done on this or other air pollution topics.

Groups will need pieces of butcher paper or poster paper, felt markers, old magazines, and glue.

An alternative method of pursuing this exercise would be to assign the topic of "ozone" to students before they construct their poster. Student groups could then use their specific research information for their posters.

Grading this assignment can be done in two ways:

1. You could grade the posters as the groups hold them up to the class.
2. You could hang all the posters around the room, have the groups rate each one, then average the ratings to assign the final grade.

In either case, we recommend using some criteria like the following:

1. a written slogan (20%)
2. a drawing or other picture (20%)
3. lots of color (20%)
4. The poster has a *warning* or *provides information* about the ozone layer and the potential dangers if it gets too thin. (20%)
5. general impression or impact of the poster (20%)

Space Science: Analysis

18. Problems in Space Travel

Teacher Instructions

This exercise is designed to provide students with an ongoing project during any study of space. It can last as long as you want it to, or it could be done in three days: Day 1—thinking of problems; Day 2—group problem organization and selection of problem to research; Day 3—presentation of solutions to the problems.

The Space Test is intended to be a culminating exercise, but it could be used alone or not used at all. It is designed as a group test. Three people to a group works nicely. You should get some spirited discussion, especially if you allow the groups to do the extra-credit portion of the test.

Here are some problems you might wish to be sure are included on a student's list.

- At speeds we can currently achieve, any voyage outside of our solar system would take hundreds of years. On long space voyages, how would you train replacements for crew members like doctors and navigators who die?
- How would you deal with lawbreakers on a space voyage?
- On a four- or five-year trip to Mars, how would you feed the voyagers?
- At speeds we can currently achieve, any voyage outside of our solar system would take hundreds of years. On long space voyages, how would you guarantee enough genetic variability among the crew members to maintain a viable population?
- How would you counteract the effects of weightlessness on a voyage of several months or years?

Answers to Space Test Questions

1. a. Answers will vary. Be sure each pod is labeled.
   b. Answers will vary. Look for a description of what goes on in each pod.
   c. Answers will vary. Needed in this answer is an explanation of how what is going on in the pod solves one of the problems from the list.
2. Answers will vary. Check to see that the way they connect the pods with tubes shows how materials would cycle.
3. a. Answers will vary. The best way to get this type of ship out of the atmosphere would be to send separate parts up to a space assembly station. The completed ship would then not have to deal with the problem of friction in the atmosphere.
   b. Answers will vary. Be sure that the type of power suggested is feasible with today's technology. "Impulse engines," "warp drives," and "hyper drives" are really not acceptable, unless you want the students to invent some new form of propulsion. This question is designed to take advantage of the expertise of the group information and presentations from the previous days.

Scribe's Name ______________________________ Date ____________________

Space Science: Analysis

Space Test

Time Limit: Forty Minutes

Get into groups as directed by your teacher. Carefully read this entire test before you begin.

Below is a drawing of a prototype spacecraft designed for an extended flight to the outer planets. Your group's job is to label as many of the pods as you wish with specific functions designed to help you solve the problems on the list below the diagram.

On your answer sheet, draw your own diagram of the spacecraft and:

1. a. Label the pods

   b. Describe what goes on in each one, and

   c. Explain how it solves one of the problems from the list.

2. Connect the pods with tubes you draw yourself to show how materials would cycle, as you work on the problems.

3. a. Explain how you got the parts of the spacecraft out of Earth's atmosphere, and

   b. Explain how you plan on powering the ship while in space.

You may choose to block off or isolate any of the pods, or enlarge or shrink a pod. You may connect the pods in any way you wish. They do *not* all have to connect to the main cabin; in other words, you may erase some of the existing connections. The bottom line is: You have six pods and one main cabin to work with.

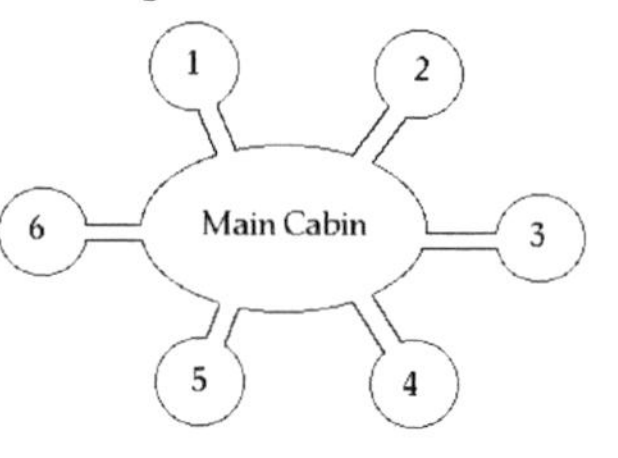

Problem List:

solid human waste	liquid human waste
air	solid nonhuman waste
water	(problem of your own)

For Extra Credit: After you have completed all portions of this exam, you may explain how your group would deal with the following emergency while on your space journey:

While working outside of the spacecraft, one of the mechanics loses his or her air supply for ten minutes. Back in the ship, the mechanic is revived, but remains in a coma. There is no evidence of brain activity. Breathing and heart rate continue normally.

Space Science: Creative Writing

19. Death of the Sun

Teacher Instructions

This is a good assignment to use with the procedural suggestions on pages *ix–x* of this book.

With the underlining and ALL CAPITAL LETTERS instruction for the important words in this exercise, the grading of the story becomes fairly simple. You could even provide the correct sequence to cooperative groups and check the order during or after the reading of the stories in the group.

Most earth/space science textbooks have more than enough information needed to complete this writing assignment.

The following is an outline is one possible sequence for the story.

A Typical Scenario for Death of the Sun

- As the sun begins to die, it will first expand to become a RED GIANT. It will then be very much larger, redder, and closer to Earth. (The sky above will be different.)
- The Earth's temperature will rise dramatically and the polar ice caps will begin to melt, leading to large-scale flooding. Islands and coastal areas will begin to disappear.
- The Earth's surface will be scorched, crops will die, and farmland will become baked and useless.
- The Earth's oceans will turn to steam, and all water vapor will be burned out of the atmosphere.
- The Earth's atmosphere will be driven off into deep space.
- The red giant will have continued to expand until it engulfs its nearest neighbors, Mercury and Venus.
- By this time all human life (all life) would have perished, unless humans had mastered interstellar travel and had escaped to another solar system.

Answers to Questions (In-Class Student Follow-up)

1. 5000 million years
2. 5000 million years
3. It will expand and become bigger and redder.
4. Mercury and Venus.

5. It will be driven off into deep space
6. Melting of the polar ice caps
7. Red giant

Correct Sequence for the Word List

Only the following words from the list have a definite sequence. All of the other words should just be graded as present, if included correctly. EXPAND, RED GIANT, TEMPERATURE, POLAR ICE, MELT, FLOODING, STEAM.

Space Science: Creative Writing

20. An Interview with Galileo

Teacher Instructions

On the day the assignment is due, follow the student instructions for the in-class portion of the assignment. It is always a good idea to collect the stories when the students arrive and stamp or mark them to show on-time completion of the assignment. Return the stories while the students are working on the consensus questions.

Answers to Questions

1. As a boy, Galileo showed interest in music (playing the lute and the organ), in building toys, and in painting. At a later age he also studied medicine and philosophy.
2. He observed a lamp swinging back and forth. This gave him the idea for formulating the laws that govern the movement of a pendulum. He then used a pendulum to time the pulse rate of medical patients.
3. Galileo invented a hydrostatic balance (for measuring the weight of objects in water) and the sector (a type of compass used by draftsmen). He did *not* invent the telescope.
4. Galileo reasoned that gravity would pull all objects downward at the same rate, regardless of their weight. He showed this was in fact true when he dropped a ten-pound weight and a one-pound weight from a tower simultaneously and they both hit the ground at the same time. Unfortunately, many people still believed in the teachings of Aristotle, who said that the ten-pound weight would fall ten times as fast as the one-pound weight. This experiment so angered many of Galileo's colleagues and some church officials that he was relieved of his teaching position at the University of Pisa.
5. He discovered that the moon was covered with valleys and mountains, and that it reflected light from the sun. Again, he was in opposition to the followers of Aristotle, who had said the moon was a smooth sphere shining with its own light.

6. With an improved telescope, Galileo observed the Milky Way and found that it was made up of stars "so numerous as to be almost beyond belief."
7. He discovered the four moons of Jupiter.
8. Much of his work supported the theories of Copernicus.
9. The Catholic Church held the belief that the universe was Earth-centered. The Church warned Galileo that his theory of a sun-centered universe was unacceptable, as was any other work which tended to support that theory. After the publication of his masterpiece *A Dialogue on the Two Principal Systems of the World,* he was called before the Inquisition and eventually forced to say that he no longer believed in a sun-centered universe. He was also sentenced to an indefinite period of imprisonment, but the sentence was not enforced.
10. *Dialogues on the Two New Sciences* was a summary of all the work Galileo had done in the study of motion, acceleration, and gravity.
11. The *Dialogues* served as a basis for the three laws of motion proposed by Isaac Newton.
12. Galileo spent the last few years of his life in his own home under the observation of members of the Inquisition. He had gone blind several years earlier.

Space Science: Creative Writing

21. Story of a Star

Teacher Notes

This is a good assignment to use with the procedural suggestions in the front of this book.

With the underlining and ALL CAPITAL LETTERS instruction for the important words in this exercise, the grading of the story becomes fairly simple. You could even provide the correct sequence of events in the life of a star to cooperative groups and check the order during or after the groups' reading of the stories.

Most earth/space science textbooks have more than enough information needed to complete this writing assignment.

Students will need to decide whether they began their life as a relatively low-mass star, like our sun, or a massive star. Their choice determines how their life story ends.

An outline of one possible sequence for the story follows.

Typical Sequence of a Star's Life

- Gravity and dust in a nebula are pulled together by gravity.
- Nebula shrinks and forms in the shape of a ball.
- Shrinking causes temperature to increase to millions of degrees.
- Hydrogen atoms begin to fuse to create helium (nuclear fusion).
- Even greater temperatures and light are created by fusion and the star begins to shine (the birth).
- Shrinking (due to gravity) stops and the star will shine steadily for millions of years until hydrogen fuel is depleted.
- When the fuel is depleted, the star begins to collapse, again due to gravity, and then begins to heat up again.
- Increased heat leads to new nuclear reactions that cause the star to expand and become a red giant.
- At this stage, there are two different scenarios: (1) A medium-sized red giant will stay a red giant for millions of years, gradually shrinking and eventually becoming a very dense white dwarf; (2) a very massive red giant will continue to expand until it becomes a super giant.
- A super giant will eventually run low on fuel and start to collapse; this leads to a massive explosion and what is identified on earth as a supernova.
- After a supernova, what is left of the star collapses into a tiny ball of neutrons, even smaller and denser than a white dwarf. It may start to rotate and give off radio waves. Then it is called a pulsar. Very massive stars continue to collapse until there is no matter left, only a tremendous gravitational pull that even attracts light. This is known as a black hole.

Answers to Questions

1. Vast cloud of dust and gas
2. White dwarf
3. Thousands of millions (billions) of years
4. Nuclear fusion
5. 2 hydrogen atoms $\rightarrow$ 1 helium
6. Gets higher
7. Nebula, star, red dwarf, white dwarf, supernova, pulsar, black hole

Correct Sequence for the Word List

If the star begins with a very large mass:

Only the words in the following list have a definite sequence. All of the other words should just be graded as present, if included correctly. NEBULA, YELLOW, RED GIANT, WHITE DWARF, SUPERNOVA, PULSAR, BLACK HOLE

If the star begins with a relatively small mass:

Only the words in the following list have a definite sequence. All of the other words should just be graded as present, if included correctly. NEBULA, YELLOW, RED GIANT, WHITE DWARF, BLACK HOLE

Space Science: Analysis

22. Lost in Space

An Activity in Recognizing and Analyzing Relevant Data

Teacher Instructions

The Location of *Discovery* is narrowed down to the two planets closer to our sun than Earth by the statement "The sun appears to be larger in the sky than you remember it to be when on Earth." The search for relevant information then should concentrate on Mercury and Venus.

The entry "Verification of Location Based On:" (line 11.0) is where students should list the references they used in determining their location.

Answers to Questions

1.0. **Log Entry 2.0** is the student's name; entries **3.0–5.0** may include any reasonable sounding information.

6.1. **Galaxy:** Milky Way

6.2. **Star System:** The intention here is that students will name this the solar system or say "the star system containing the planet Earth."

6.3. **Planet:** Venus

7.0. **Observations that support planet location with explanation include:**

7.1A. Rocky terrain with rolling hills

7.1B. Rocks glow red.

7.1C. Sun appears to set in the east.

7.1D. The sun appears to have moved only slightly in the sky.

7.1E. Pale yellow clouds

7.1F. Clear white sky, view of planetary neighbors is obscured

7.2A. Consistent with known topographic features of Venus discovered by recent spacecraft orbiting the planet

7.2B. Surface temperature is over 800 degrees Fahrenheit.

7.2C. Venus rotates in the opposite direction from Earth.

7.2D. Venus's period of rotation is 243 days, so one Venus day is equivalent to approximately 121 earth days.

7.2E. Venus is known to have clouds composed of sulfuric acid droplets.

7.2F. Earth is unable to see the surface of Venus due to its extremely heavy cloud cover. (This is also the reason Venus glows so brightly in our night time sky; clouds reflect a great deal of sunlight.)

8.0. **Approximate Distance from Center of Star System:** 67 million miles or 108 million kilometers.

9.0. **Approximate Diameter of Planet:** 7250 miles or 12,100 kilometers

10.0. **Nearest Planetary Neighbor in Star System:** Earth. (This answer may require students to subtract Venus's distance from the sun from Earth's distance from the sun, and also Mercury's distance from the sun from Venus's distance from the sun, in order to determine Venus's closest neighbor.)

11.0. Check for proper annotation.

12.0. This section allows students to put down more information from their research without structuring the data. Accept any reasonable information about the planet.

Space Science: Analysis

23. How Many *Elis* Tall Are You?

Equivalency Assignments

Teacher Instructions

There are several ways this assignment can be used. The original design was to familiarize students with the concept of equivalencies with the *Eli* Lab before doing the astronomical conversions. The astronomical conversions could be done alone, as well. This is why the Earth Equivalency assignment begins with the same introductory paragraph as the Personal Equivalency assignment.

For the How Many *Elis* . . . exercise, the following choices of student standards (step 3) are recommended: for height—the shortest person in the class; for head size—the largest person in the class; for distance from the front—someone in the center of the room; for foot size—an average-sized female. These selections are not arbitrary. By choosing the shortest, largest, middle, and average person for each standard, students will have to compute their sizes in more than one standard unit (height), and less than one standard unit (head size), and will have two other sizes that might be greater than or less than one.

You could have students do some organizing of data by having them graph their height in *Elis* (x-axis) vs. any of the other equivalents on the y-axis.

The chart below lists the Earth equivalents for each planet and each characteristic.

Planet	Distance from Sun	Diameter	Revolution	Rotation
Mercury	.39	.38	.24	58.66
Venus	.72	.95	.62	243
Earth	1.0	1.0	1.0	1.0
Mars	1.52	.53	1.88	1.04
Jupiter	5.19	11.20	11.86	.42
Saturn	9.51	9.41	29.46	.46
Uranus	19.13	4.01	84.01	.71
Neptune	30.00	3.87	164.79	.66
Pluto	39.33	.18	247.69	6.36

An excellent follow-up assignment is to have student groups produce solar system diagrams on butcher paper using various scales (i.e. earth–sun = 1 cm; earth–sun = 1 dm, earth–sun = 1 m). To do the earth–sun = 1 m diagram, students will have to go outside and place labels where the planets would be in a long line. Borrow the tape measure that the school track program uses for measuring discus throws to measure the necessary 40 meters on the outside scale.

You could also chart the planet diameters in a similar way. If the students start one edge of each planet at a common point as shown below, they can get all of the planets on one sheet in the 1 cm scale, four sheets of fan-fold paper at the 10 cm scale, and we recommend going outside to do the 1 m scale. In the diagram, *p*1, *p*2, and *p*3 stand for planet edges.

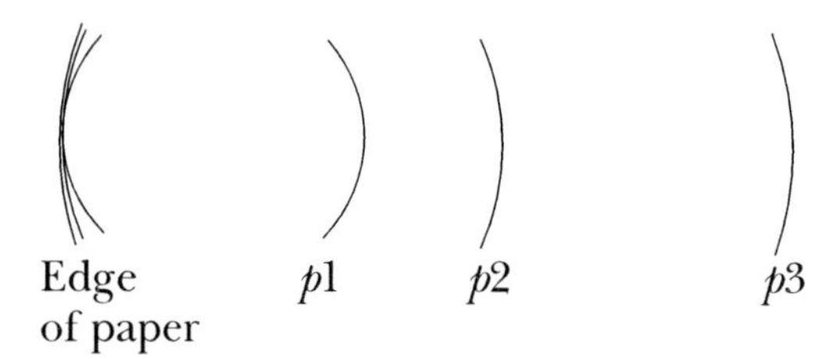

Space Science: Art

24. A Trip to My Favorite Planet

Teacher Instructions

This could be a take-home assignment or an individual or cooperative assignment in class. If you choose to do this as a cooperative, in-class assignment, the following guidelines are recommended:

- Have students work in pairs or trios.
- Allow students only a portion of the period for completion of this task. Twenty five minutes should be adequate.
- Establish accountability by requiring the entire group to show its poster to the class.

If this assignment is done in class, you will need to provide colored pencils or pens and paper.

We recommend limiting the number of students who choose any single planet in each class. You could include the asteroid belt, the sun, and the moons of the various planets as travel destinations if you wanted to. You might also want to have students sign up for the planet they wish to do their poster on.

Grading this assignment is fairly subjective. We give 100% to outstanding posters. We award an A grade to any poster that was produced following the instructions completely. Major deductions should be made for:

- no color
- incorrect number of slogans
- incorrect style of slogans
- destination not given or in wrong location
- wrong size

You can use this as a way to give extra credit if you are so inclined. Simply lower the possible value of the assignment by 40%. Award the revised full credit amount to anyone who follows the instructions. Award up to 40% extra credit for the best, most creative posters you get.

It is very easy to add a short written portion to this exercise by assigning a one-paragraph description of the planet, to be attached to the back of the poster.

Space Science: Art

25. "The Creature That Came from . . ."

Teacher Instructions

The major purpose of this assignment is to provide students with an opportunity to do some artwork for the class. While the writing is important, we recommend weighting the final grade toward the quality of the illustration.

Groups in this exercise could be as large as five students, with each group member choosing one of the planets listed.

We recommend allowing students three to five days to research, draw, and write.

There are at least two ways to group the students for the in-class portion of the assignment. We do not recommend either one over the other. The two ways are:

1. in their original groups from the first day. This option allows all students to hear and see ideas from each planet.
2. in groups by planet. This option shows students the variety possible in designing something for a specific situation.

Answers to Questions

Most of the questions are designed to stimulate discussion within the group. They are intentionally open-ended. The process of reaching consensus is as important as a reasonable answer.

a. Answers will vary. Look for some way to dissipate heat.
b. Answers will vary. Look for some way to conserve or generate heat.
c. Answers will vary. Students could provide a defense for no differences, but considering the lack of oxygen on Uranus, there should be some difference discussed.
d. Answers will vary. Look for some way to keep close to the ground, like extra weight. A Martian of the same build as a human could easily jump 30–40 feet. Really perceptive students could argue that the lesser gravity

would make no difference, since the creature would have grown up on that planet and would not know it had less gravity than anywhere else.

e. Answers will vary. We don't have a clue. Gases?

f. Answers will vary. Look for something for heat, pressure, or extended light or darkness.

Physical Science: Creative Writing

26. That Really Burns Me Up

Teacher Instructions

This assignment is designed as a true, cooperative jigsaw assignment.

This assignment was developed for groups of four students, but smaller groups could be accommodated in two ways: (1) one group member could research and write two portions of the group dialogue, or (2) you could eliminate one of the research areas or allow the group to choose which area they would not do. Individual students could research and write dialog for all four parts, but this should then be assigned as a long-term assignment.

Groups larger than four students could be accommodated by having students double up on parts. On the second day, the subgroups could combine their dialogue information or select one of the two student options to use in the group dialogue. If you choose to use larger groups, make certain that all of the dialogues are read aloud and turned in for their individual grade.

The purpose of this assignment is to expose students to a large part of an environmental problem. It is possible for groups to get slightly different information on any of the four parts. This is most likely in parts 1 and 4, which deal with the nature of ultraviolet light and the damage it causes. Differences in information in parts 2 and 3 are more likely to be in degree of coverage.

You may need to allow more time for in-class research for step 5 of the In-Class Student Follow-up. If students decide something is missing, they probably do not already have the information necessary to write new dialogue and will need time to do more research.

Questions i and j are not part of the student research requirement. They are included to stimulate thought and discussion.

Answers to Questions (3.a.–j.)

a. **Adenine** (**A**), **guanine** (**G**), **thymine** (**T**), and **cytosine** (**C**) are the four nitrogenous bases in DNA. They form the "rungs" in the DNA ladder. **A** pairs with **T**, and **G** pairs with **C**.

b. The **dermis** is the middle layer of the skin. It lies below the epidermis. The dermis is the *true* skin and contains nerve endings, blood vessels, hair follicles, sweat glands, and connective tissue.

c. Different forms of radiant energy are distinguished by their wavelength. The **electromagnetic spectrum** is the combined listing of these energy forms. The wavelengths in the electromagnetic spectrum range from very short, high-energy gamma rays to very long, lower-energy radio waves.

d. Skin turns red when exposure to ultraviolet light reaches the first stage of skin damage.

e. The germinative layer of cells tries to protect the lower layers of skin by causing the melanocytes to produce extra melanin pigment. Melanin absorbs ultraviolet radiation.

f. Ultraviolet radiation damages DNA. It causes mutations. The actual damage caused involves consecutive thymine bases in DNA. When ultraviolet light strikes a portion of DNA with T-T in the sequence, the two thymines fuse. The thymine dimer (compound) that is formed reads as a single base, usually guanine, and changes the reading sequence of enzymes on that sequence. Damage to collagen is also caused by ultraviolet radiation. Damaged collagen loses elasticity, and wrinkles form.

g. Normal, undamaged DNA is responsible for the correct functioning of the cell. It directs production of proteins and is responsible for passing inherited traits on to future generations of cells.

h. Skin cancer is the result of multiple damages. Skin cells damaged by ultraviolet radiation lose their ability to recognize other cells. Instead of dividing only until they contact other cells, they continue to divide uncontrolled. A cancerous tumor is formed.

i. Decreasing the ozone layer, which protects the earth from certain forms of radiation, will allow a higher intensity of ultraviolet radiation to reach the earth's surface. More ultraviolet radiation at the surface increases the chances for DNA damage.

j. There is no correct answer to this question. In general, the higher the SPF value, the greater the protection. Total blocking of ultraviolet light might actually be the best thing for the skin, although studies are still being done.

Physical Science: Analysis

27. One Step at a Time

A Demonstration of Enzyme Function and Enzymatic Pathways

Teacher Instructions

You will need scissors and glue for the students to use in this exercise.

This exercise does not take a full class period. Twenty to thirty minutes should be sufficient for completion of this task.

You should copy the pages that follow on the color paper recommended: cone = tan or buff; ice cream = white; cherry = pink or red; sign = blue; price = yellow or goldenrod. You will need one of each part for each group.

After printing, cut the individual pieces into rectangles to hand out to the groups.

The exercise was designed to have the individual parts glued down to the background in this order: cone—ice cream—sign—cherry—price. We recommend that you make a sample poster in this order. See the sample on the next page.

In order for the simulation of an enzyme pathway to be fully effective, you must demand that the students follow the instructions explicitly.

Answers to Questions

1. Answers will vary—these are names and jobs.
2. Answers will vary—this is a list of gluing order.
3. Answers will vary—it should be a time.
4. Answers will vary—it should be a time.
5. Answers will vary—it should be a time.
6. Answers will vary—there should be some explanation for whichever step they select.
7. Answers will vary—if "no," there should be an explanation and a new gluing order. Hopefully it will be identical to the order in the teacher note above.
8. Answers will vary. Probably the poster would have had some part incorrectly glued on top of another.
9. Answers will vary. The designed limiting factor is the cutting of the price star, but gluing is a possibility, too. In any case, there should be some suggestion as to how to speed up the process.

One Step at a Time

Sample Poster

One Step at a Time

Templates for the poster

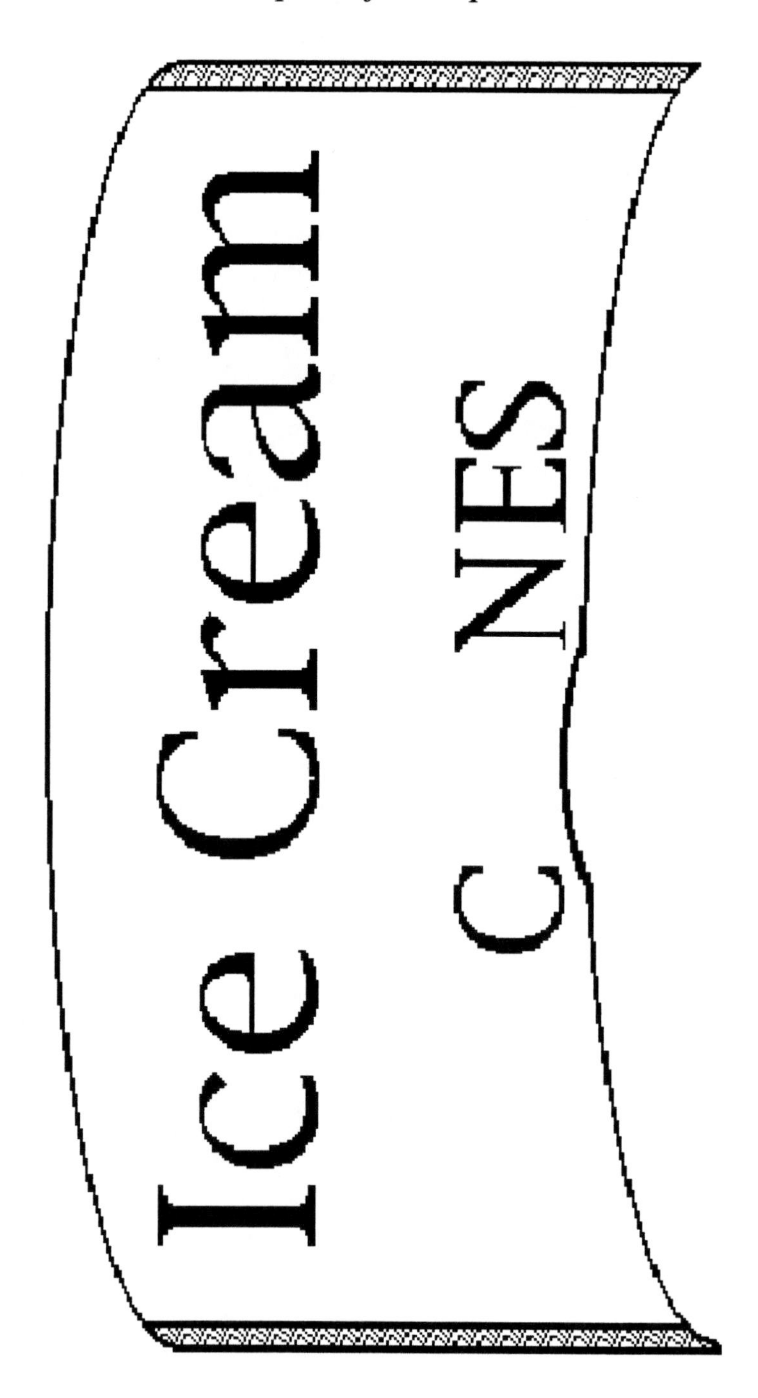

One Step at a Time

Templates for the poster

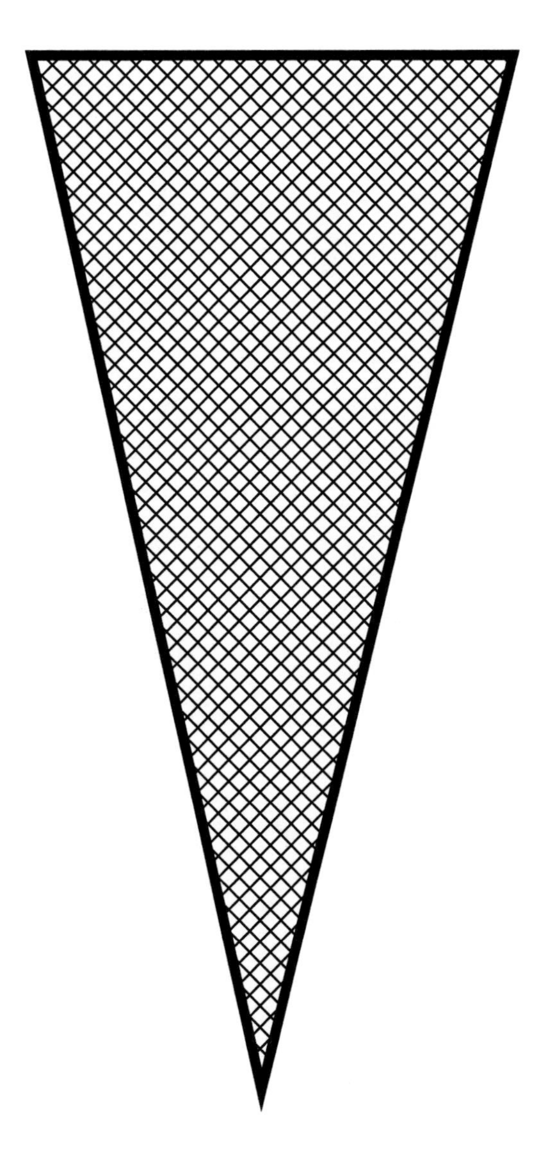

One Step at a Time

Templates for the poster

Physical Science: Laboratory Skills

28. Liver Without Onions in a Peroxide Sauce—Yum!

Teacher Instructions

The lab can be used to illustrate enzyme action, digestion, or an exothermic reaction.

You will need one or two large test tubes, one long thermometer, hydrogen peroxide, and some liver for each group of students. One test-tube rack per group is a good idea as well. A good group size is two to three students per group.

One pound of liver will do fifty to sixty groups. Cut the liver into roughly .5 cm chunks.

Be sure to answer any student questions written down on the prelab before beginning the laboratory exercise.

The liver foams and bubbles when it contacts the peroxide. The foam usually oozes out of the test tube and is pretty gross. If you kind of act like that process is a surprise to you, and that it could mean that somebody goofed up, it might help focus the students who would otherwise tend to loudly "gross out" at the foam.

Caution students to be careful when inserting the thermometer into the test tubes. Thermometers with glass bulbs can break or push through the bottom of the tube, especially once the reaction begins.

Answers to Questions

1. Answers will vary. Foaming, temperature change of the peroxide, and color change of the liver are all possible answers.
2. Answers will vary. The operative word in the question is *think*, so this allows for some latitude in the answers. "Oxygen is released as peroxidase breaks down the H_2O_2" is the correct answer. Accepting the release of hydrogen gas is probably stretching the acceptable limits for this question.
3. Answers will vary. The liver turns a gray color. The texture changes to much less firm. Another interpretation of the question would produce an answer like "enzymes acted on something in the liver."
4. Answers will vary. Probably not. The reasons for differences in change of temperature over successive time periods include using up of the substrate.
5. Answers will vary. Check the charts to see that the answer matches the time period with the greatest variation.

6. Answers will vary. Probably. The data tends to be fairly consistent.
7. The release of a gas from the reaction. Specifically, the release of oxygen.
8. To verify the reaction's progression. To validate the results.
9. Answers will vary. The general trend would be the same; however, the chances of the data from any one run being equal to the average, while not nonexistent, are slim.

Physical Science: Laboratory Skills

29. Funnels of Fun

Teacher Instructions

This is an effective way of demonstrating the concept of limiting factor while allowing the students to participate in an activity that is fun.

You need a large selection of funnels of various sizes. There is no need for each group to have the same sizes or shapes of funnels, so just grab all the funnels you can find.

It facilitates the process if each student group has a small bucket for catching the water poured through the funnels. Beakers of 500–1000 ml size would work for this purpose, but breakage could be a problem. If you *do* use beakers to catch the water, students could reuse most of the water they used for the first pour on consecutive trials.

The number of times the funnels are rearranged is up to you. Three times, as recommended in the procedure, should be enough to demonstrate the principle without leading to water fights, etc.

The graph suggests the *y*-axis for time because time is the dependent variable in this exercise.

Student groups of five or six work well in this exercise.

Answers to Questions

1. Answers will vary.
2. a. Answers will vary. The answers should be names of the group members in the **Name** column of the chart they are instructed to make.

   b. Answers will vary. The answers should be in seconds. There should be one answer for each group member in the **Time** column of the chart from part **a**.

c. Answers will vary. The answers should be in cm or mm. There should be two answers for each group member, and the answers should be in the **B.E.** and **S.E.** columns of the table.

3. Answers will vary. The answers should be in the form of a graph. Check for the "time" to be the *x*-axis. Students can make two graphs, one for each end of their funnels, or one graph with two colors, one for each end of their funnels. The general relationship is the larger the small end of the funnel, the faster the water runs through.
4. a. Answers will vary. This should be a list of names.

   b. Answers will vary. This should be a different list of names. By this time, students should be noticing that the time is very similar, or the same, regardless of the arrangement of the funnels.
5. This answer should be the funnel with the smallest diameter.
6. The limiting factor is always that part of the reaction that determines the time for the reaction to occur. In this demonstration, the smallest funnel is the limiting factor.

Physical Science: Laboratory Skills

30. Sometimes You Feel Like a Nut

Teacher Instructions

You will need to provide four or five different kinds of nuts, including a dry-roasted peanut and a regular peanut, roasted in oil. Other types of nuts that burn well are pecans, walnuts, Brazil nuts, and almonds.

Equipment needed per setup includes one each:

nut
cork
straight pin
250 ml Erlenmeyer flask with 100 ml water
ring stand
clamp
2-pound coffee can (with both top and bottom removed and a V-shape cut in one edge)
thermometer
single-hole rubber stopper that fits both the flask and the thermometer
matches

In student step 7, the 50 ml recommended assumes 125 ml Erlenmeyer flasks. If you use 250 ml flasks, you can use 100 ml of water. This simplifies the math, too.

In the formula in question 4, 1 calorie per gram of water is a constant. Be sure to inform your students of this or they will probably end up multiplying by 50 or 100, depending on how much water is used in the flask.

Provide one demonstration setup as a model for student setups. See the following diagram for setup of equipment.

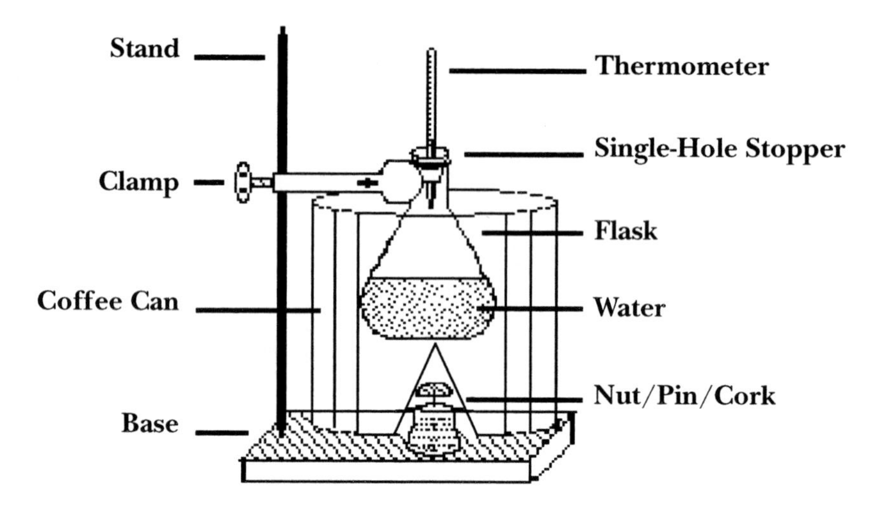

The nut must be stuck on the tip of the pin to burn properly; it will not burn well in a dish, and it may set the cork on fire if the two are touching.

Answers to Questions

1. Oil or fat
2. Insulation or concentration of the heat
3. Answers will vary. The calculation should show "Temp After Nut Burned" – "Temp Before Nut Burned" = "Net Change in Temp."
4. Answers will vary. Calculations should follow the form given in the student questions:

   (Mass of Water) × (Temp Change of Water) ×
   1 calorie per gram of water = Calories absorbed

5. Answers will vary. This calculation is dividing the answer to question 4 by 1000 to convert to food calories (kilocalories).
6. Answers will vary. This calculation should be the answer to question 4 divided by the weight of the nut used. The answer should be in calories per gram.
7. Within each class, all team answers should be the same.
8. Within each class, all team answers should be the same.
9. Within each class, all team answers should be the same.

10. A regular peanut will have more calories than a dry-roasted peanut because the regular peanut has been roasted in high-calorie oil or fat.

Physical Science: Laboratory Skills

31. Faster Than a Speeding . . . Snail?

Teacher Instructions

While this exercise uses an animal, it is not necessarily a biology exercise. Students collect a variety of data and do a number of different observations. Observation, data collection, and data analysis are the primary goals of this exercise.

Students can work in pairs or individually.

When all of the students are finished, analysis of class data can be done. Students could produce a frequency chart for snail length or number of breaths, or graph snail length vs. weight. Another type of analysis is graphing weight vs. speed. As a crossover into physics, students could calculate the amount of work done by a snail. For example:

$$F = ma$$

$$a = \frac{v_1 - v_0}{t_1 - t_0}$$

$$W = F_s \cos\emptyset$$

Since the snail is moving in the same direction as the force it is exerting, the formula becomes

$$\text{Work} = \text{Force} \times \text{distance because } \cos \emptyset = 1$$

If a 50 g snail sped over its 15 cm race course in 15 seconds:

$$F = 50g \frac{1\ cm/\text{sec} - 0\ cm/\text{sec}}{15\ \text{sec} - 0\ \text{sec}} = \frac{50g - cm/\text{sec}^2}{15} = 3.33g - cm/\text{sec}^2$$

Work done = 3.33 g-cm/sec^2 × 15cm = 3.33 g-cm^2/sec^2 = 50 ergs

Unless you have a large number of scales, the weighing of the snails will be the limiting factor in this exercise. Students can skip that step and do it anytime one of the scales is not in use.

Answers to Questions

1A. Answers will vary. This answer should be in grams.

1C. Answers will vary. This answer should be in cm or mm.

2C. Answers will vary. However they all should mention the "foot" that can be seen undulating by looking through the glass. Some mention might also be made of the slime.

3A. Both sexes in the same individual. Hermaphroditic also known as monoecious.

3C. Humans have individual sexes.

4C. Answers will vary. Students should have three numbers, one for each trial. Numbers should be separated by commas.

4F. Answers will vary. This should be the average of the numbers in 4C.

4G. Answers will vary. This formula should be filled in with the answer to 4E.

$$\frac{15\,cm}{?\text{sec}} \times \frac{3600\,\text{sec}}{1\,hr} \times \frac{1\,m}{100\,cm} = \text{the answer in mph}$$

4H. Answers will vary. This answer should be one of these: **faster**, **slower**, or **the same**.

5B. Answers will vary. This should include something about the scraping of the lettuce or the term **radula**.

6A. Snails have four antennae.

6B. The back, rear, or second ones. Answers will vary as to why. The *real* reason is that is where the eyes are located.

6C. On the longest or rear tentacles or antennae. The snail protects its eyes by retracting the tentacle (antenna) when it is touched.

6D. Answers will vary. Something about antennae movement should be included.

7A. Answers will vary. This should be a number.

7B. Answers will vary. This should be a comparison of the number in answer 7A to the number of breaths a human takes in one minute (17). Usually the snails breathe faster.

Physical Science: Creative Writing

32. Eddie (or Edie) Electron

Teacher Instructions

This assignment allows the students some creative license as they trace the path electrons follow through a circuit. Encourage students to write in the first person and describe their feelings as they encounter different amounts of resistance at different points in the circuit.

This story could be done as a jigsaw activity in a group of four students during one class period. Each student would research two definitions independently, then the group of four would come together to put all of the information into story form. There would be some overlap in information gathered by students, but this should be an aid in understanding circuitry and writing the story.

Definitions

battery —provides a difference in potential to make the current (electrons) flow. (The negative terminal has an excess of electrons, and the positive terminal has a deficiency of electrons, so current flows from negative to positive. The electrons are "pushed" through the circuit by this difference in potential.)

negative terminal—has an excess of electrons. The electrons start in the circuit here and are "pushed" by the difference in potential.

positive terminal—has a deficiency of electrons. The current moves toward this terminal.

insulation—a material with low conductivity and high resistance. Insulation is used to coat wire so that current does not flow out of the intended pathway.

wires—provide a pathway for electron flow. Ideally, they are made of a metal, are good conductors (have low resistance), and cover the shortest pathway possible.

resistance—an obstacle to the flow of electrons. Materials with a high resistance are not good conductors and are therefore used to insulate or block the flow of current.

light bulb (a source of resistance)—the filament in the light bulb provides a definite amount of resistance. Since it is part of the pathway that electrons must follow and has high resistance to electron flow, heat is built up and the filament begins to glow, producing light.

switch—turns the current on or off by completing or breaking the circuit pathway. A switch contains a material that is a good conductor. When a switch is closed, current flows from the wire, through the switch, and back into the wire on the other side. If the switch is open, there is a gap in the circuit and electron flow is stopped.

Order of Story

negative terminal of battery—Eddie is with many of his friends here and they feel a push, moving them out of the battery and into the wire.

insulated wire—travel is fast and easy along the wire with little heat built up, but it's very difficult to leave the pathway and pass through the insulation.

switch—when the switch is closed, travel continues at the normal speed, but if the switch is open, then everyone comes to a complete stop.

wire—see above.

light bulb—electrons are still at top speed as they move into the light bulb, but when they reach the filament, traffic begins to back up. It's much more difficult to move through the poor conductor and everyone moves very slowly. A lot of heat is built up, and as a result, the filament begins to glow.

wire—it's out into the wire again, and as speed picks up there is very little resistance, and therefore very little heat built up.

positive terminal of battery—Eddie reaches the end of the circuit; there are not so many electrons here as at the start.

Answers to In-Class Group Questions

1. Good conductors would include copper, aluminum, any other metals, water (that is not distilled), some types of ceramics, and carbon (dust or electrical brushes).
2. Good insulators would include glass, plastic, rubber, and porcelain.
3. This question is meant to heighten students' awareness that batteries can be made of many different materials and for many different purposes. Some types of batteries are nickel-cadmium, gel, and alkaline.
4. A light bulb glows because its filament is a poor conductor of electricity; as electrons meet the high resistance of the filament, their energy is converted into alternate forms, namely heat and light.
5. The greater the distance an electric current travels, the greater the heat build up due to resistance. Thin wire extension cords are designed for short distances so that little heat should build up. Therefore, if you connect several short, thin wire extension cords end-to-end, the greater length of the wire will increase resistance and heat buildup and may result in overheating the cords and starting a fire.

Physical Science: Laboratory Skills

33. Fish Respiration: This Lab's All Wet!

Teacher Instructions

If you and your students follow all of the guidelines, you should not lose a fish during this procedure. However, since we recommend using feeder fish (see below), you might lose one or two because they might not have been very healthy to begin with.

This is an excellent data collection lab. While it can be used to illustrate how poikilothermic (ectothermic or exothermic or, if you absolutely have to, cold-blooded) organisms are affected by temperature, it does not have to be used in a biology class. This would be a good way to integrate some life science into a physical science class when discussing dissolved gases.

Feeder goldfish from a local pet store work fine. You should be able to get eight to twelve for $1.00. You will need two complete sets of fish if you do this lab exercise all day. Rotate the fish period by period. This keeps the trauma on the fish to a minimum. We found that we could sell the fish to students as pets for $0.25 each the day after the lab, since some of the kids form an attachment to their fish during the procedure.

If you use full-size kitchen ice cubes, you will need only one cube per beaker to get a nice result. You will have to experiment with smaller cubes ahead of time. You need to keep the temperature within the range highlighted in the student instructions, so do not add too many cubes of any size.

The warm water added to the beakers should be about 50 degrees C. Again, observe the temperature limits. If you want, you can add warm water a second time after step 13. This is not necessary unless the room you are in is very cold and the beakers lose enough heat to get back to room temperature by that time.

Student groups can be pairs or trios. Individual students sometimes have a hard time doing all of the various readings and recordings. Also, observing the gill movements in a fish is not all that easy, so two people counting is a good idea. We recommend looking straight down into the beaker to count the gill movements.

Answers to At-Home Assignments

1. This is a graph of three sets of data. Check to see if the graph is laid out correctly, if the spacing of the units is consistent, if the axes are labeled, if there are three lines on the graph, and if there is a key indicating which line represents which data.

2. The warmer the fish, the faster the respiration, or, the colder the fish, the slower the respiration.
3. **Ectotherm** (**exotherm**) or **poikilotherm**. (We do *not* accept **cold-blooded** in our classes.) This term means that the temperature of the organism fluctuates with the temperature of the environment. Cold tends to slow these organisms down. Heat tends to speed them up.
4. Snake—yes. Human—no. Snakes are ectotherms, like fish. Humans are endotherms and should not be affected by environmental temperature fluctuations.
5. There are several flaws. Accept any legitimate one. Flaws include, but are not limited to (1) it is hard to count gill movements, (2) we do not know that gill movement speed has anything to do with respiration rate, (3) cold water allows more oxygen to dissolve in it, so the fish might not need to breathe as rapidly in cold water.

Physical Science: Art

34. Move Over, Bill Watterson

Teacher Instructions

You should have colored pencils or felt pens and unlined paper ready for students to use in this exercise.

Grading of this Cranial Creation is subjective. Basing the grade on appropriateness seldom results in any complaints.

A good indicator of the appropriateness of the cartoon is the class reaction to the opaque projection of the cartoon onto the screen. If the students in the class do not understand the cartoon, but you do (or vice versa), it is probably still a reasonable cartoon. If neither you nor the students understand the cartoon, well . . . .

You will probably find the political cartoons vaguest.

This is a good way to do a unit review. Students have to demonstrate factual knowledge and infer appropriate relationships.

To help focus your students, you could put a list of acceptable topics on the board. Student groups could sign up for a topic before they begin their cartoon. This would allow you to regulate the number of cartoons on any one topic. It also allows you to make sure that the topic areas you want to emphasize are not neglected, without being heavy-handed.

It might also be interesting to have all of the students do their cartoons on a single topic, for instance waves, to see the breadth and depth of insight and understanding in the entire class.

Physical Science: Creative Writing

35. Move It!

Teacher Instructions

This assignment allows students some creative license as they report on some important aspects in the study of motion. Encourage students to write in the first person and to be proud of why they are the most important term.

While we designed this as a true, cooperative jigsaw activity, there are several ways to proceed with this assignment. One possible sequence of events for this assignment would be:

1. Assign the story and form the groups of four students for the activity. Collect the notebook paper with names and terms they will become.
2. Allow time for research and writing (variable, depending on your curriculum and depth of coverage you expect from students).
3. On the day the story is due, collect it from the students and check each one for completeness in following instructions.
4. Return the cover sheet with names and motion terms when the students begin to work on the group portion of the assignment.
5. Allow thirty-five minutes for students to listen to the stories and teach each other the information.
6. Give the quiz. One way to reduce the threat of the quiz is to make it a group quiz. If you do this, you might try allowing only written communication during the quiz portion of the assignment. This keeps the groups focused on *their* group members' ideas. You could be nice and allow the group time to work on the quiz after they have done their own work as individuals. This could provide a form of reward while continuing the cooperative nature of this assignment.

You could leave out the quiz and treat this assignment like another creative writing story assignment and use the suggested procedure at the beginning of this book, with students marking required information with some form of code (stars, check marks, numbers, etc.). You could think of other options, too.

Information for the writing of the story part of this assignment can be found in any physical science textbook or a book on motion.

Answers to the Quiz Questions

1. Speed
2. Acceleration
3. Direction of movement
4. Acceleration
5. Momentum
6. Speed
7. 5 km/hr upstream
8. Toward the center of the circular path
9. 8.89 m/sec (32 km/hr)
10. Go-cart A (A = 100 kg × 15 km/hr (1500 kg-km/hr) > B = 120 kg × 12 km/hr (1440 kg-km/hr))
11. Knowing only the speed of a hurricane would help predict only *when* it would hit land. Knowing the velocity would help predict *when* and *where* a hurricane will hit land.
12. It increases.

Scribe's Name ________________________________ Date ____________________

Physical Science: Creative Writing

36. Move It!

Quiz

1. The distance covered by a moving object in a unit period of time defines what?

2. The rate of change in velocity is what?

3. What do you have to add to speed to determine velocity?

4. What term can be measured in m/sec/sec or m/sec^2?

5. Multiplying mass times velocity calculates what?

6. Which of the motion terms could be measured in km/hr?

7. What would be the actual velocity of a boat moving upstream at 10 m/hr against a current of 5 km/hr?

8. If an object moves in a circular path, in which direction is the acceleration?

9. If a runner ran 400 m in 45 sec, what was the speed of the runner?

10. Which has the most momentum: go-cart A (100 kg moving at 15 m/sec) or go-cart B (120 kg moving at 12 m/sec)?

11. Why would knowing the velocity of a hurricane be more important than knowing the speed of the hurricane?

12. What happens to the momentum of a car when you fill the seats with passengers?

Scribe's Name ______________________________ Date ____________________

1. Are You What You Eat?

The old saying is "You are what you eat." Just how true is that anyway? In this exercise you should have a chance to find out, at least a little bit.

1. Get into groups as directed by your teacher.

2. Pool all of the food labels you brought into a single pile.

3. Analyze each label as follows: (Write the letter and the answer for each step in your notebook or on another sheet of paper.)

   a. Write down the name of the product.

   b. Calculate % of calories per serving in the food from **carbohydrates**, **fats**, and **proteins**. (Proteins have 3.1 Cal/g; carbohydrates have 3.8 Cal/g; and fats have 9.3 Cal/g.)

      Example: 2g of fat per serving (from label information)

      $$2\text{g}(9.3\ \text{Cal/g}) = 18.6\ \text{Cal from fat}$$

      $$\frac{18.6\ Cal}{\#\ Cal/serving} = \%\ Cal\ \text{from fat}$$

   c. List natural ingredients in the product.

   d. List artificial ingredients in the product.

   e. What was added to the product before it was canned or packaged?

   f. Why was each thing added? (What is the *purpose* of each added ingredient?)

   g. Does the product contain salt? If so, how much per serving?

   h. Does the product contain **cholesterol**? If so, how much per serving?

4. Staple each of the labels to its evaluation sheet.

Answer the following questions: (Group or individual? Ask your teacher.)

1. Which food had the most calories per serving?

(continued)

Scribe's Name ____________________ Date ______________

1. Are You What You Eat? *(continued)*

2. Which food had the highest percentage of calories from
    a. carbohydrate?
    b. fat?
    c. protein?
3. What was the most common natural ingredient?
4. What was the most common artificial ingredient?
5. List ingredients used as preservatives in these products.
6. List ingredients used as flavor enhancers in these products.
7. List ingredients used as visual enhancers in these products.
8. Which of the products do you think was the most modified from its original state?

Staple all of the analysis sheets together. Be sure each group member's name is on the top paper. Turn the entire packet in to your teacher.

Scribe's Name ______________________________ Date ____________________

2. Heavy-duty Writing Assignment

Researching a Specific Piece of Science

For this assignment, you are going to have to use the library here at school or the one in your neighborhood. The basis of the report will be a scientific article that you will choose yourself.

Here is the overall time line. You will need to add specific dates. Listen to your teacher.

Step 1: Select a scientific article to read and use as the basis of your report. You may use any article you find at home or in a library as long as it is approved by your teacher. Due date: ______

Step 2: After reading your article, decide on a related topic to research and write about. For example, if your article was titled "How a Mosquito Sucks Your Blood," you might choose to research anticoagulants (blood thinners) or some disease that is spread by a mosquito (malaria). Due date: ______

Step 3: Do some research in a library. Write down at least two other sources of information you will use in your report. You must use at least one other magazine article in your bibliography. Due date: ______

Step 4: Finish your research. Take notes on the articles you read. Make an outline of your paper. Write a rough draft. Due date: ______

Step 5: Prepare the final draft of your paper. The final paper must be typed or word processed. The report should be three pages long. Three pages is defined as over 2.5 pages and under 3.5 pages of double-spaced typing.
Due date: ______

A major part of your report (actually a separate, additional page) will be an annotated bibliography. On this page you will list your references and write a one-paragraph summary of each article beneath the bibliographical entry. All bibliography entries will be alphabetized by last name of the author. The correct format for the bibliography is:

(continued)

Scribe's Name ______________________________ Date ____________________

2. Heavy-duty Writing Assignment *(continued)*

Author. Title of Book or Magazine. "Title of the Article," Publisher, Date of Publication, Pages Used.

(A one-paragraph summary of the article or book will go here.)

Author. Title of Book or Magazine. "Title of the Article," Publisher, Date of Publication, Pages Used.

(A one-paragraph summary of the article or book will go here.)

The final report should be stapled together, with cover sheet or title page on top. There should be no report cover of any kind (plastic or cardboard, for instance).

Scribe's Name ______________________________ Date ____________________

3. Energy: The Choice Is Yours

We all need energy. Our bodies require it. Our homes and vehicles demand it. What kind of energy to use is the question in this exercise. Sometimes we do not have a choice. For example, if you drive a car, you need gasoline as the energy source . . . or do you?

Fill in the spaces below as you are instructed by your teacher.

1. List seven energy (fuel) sources.

____________________ ____________________

____________________ ____________________

____________________ ____________________

____

2. Copy the sources from the above list onto the chart below.

List of Sources	Expense	Pollution	Practicality	Overall Score
____________________	______	______	______	______
____________________	______	______	______	______
____________________	______	______	______	______
____________________	______	______	______	______
____________________	______	______	______	______
____________________	______	______	______	______
____________________	______	______	______	______

3. Rank the energy sources from 1 to 7 in the column labeled "Expense." The source with the lowest ranking (1) should be the *least* expensive. The source with the rank of 7 should be the *most* expensive.

(continued)

Scribe's Name ______________________________ Date ____________________

3. Energy: The Choice is Yours *(continued)*

4. Now rank the energy sources from 1 to 7 in the column labeled "Pollution." The source with the lowest ranking (1) should be the *least* polluting. The source with the rank of 7 should be the *most* polluting.

5. Next, rank the energy sources from 1 to 7 in the column labeled "Practicality." The source with the lowest ranking (1) should be the *most* practical. The source with the rank of 7 should be the *least* practical.

6. Finally, add all of the rankings across the row for each energy source and answer the questions assigned by your teacher.

Questions

1. Which of the energy sources had the lowest average? Which had the highest?

2. Based on the averages you calculated, which is the *best* energy source? Explain.

3. If you had one million dollars to invest in one of the energy sources on your list, which would it be, and why?

4. Which of the energy sources is the best all around for the environment? Why?

5. What specific changes in lifestyle would you as a North American consumer have to make if solar energy were ever going to become as common as fossil or hydroelectric power?

6. If the federal government decided to make one of the energy choices on your list illegal, which would be your choice to be outlawed? Defend your choice with at least two *specific* reasons.

7. If the federal government read your answer to question 6, agreed with it, and made that energy source illegal, what energy-related problems would still have to be solved?

Scribe's Name ______________________________ Date ____________________

4. Don't Get Bogged Down on This

In the newspaper or in a magazine like *National Geographic*, you might read an occasional story about a mummified body found on a mountaintop, buried in frozen tundra, or uncovered from thick layers of moss. Most of these finds are shuffled off to museums for careful examination. In the museum, modern technology might enable a scientist to determine what the person ate for a last meal, what material his clothes were made of, or whether he should have been taking vitamin pills.

Suppose that you read this in the local newspaper one summer morning:

Scientists today removed the body of a young person from the peat bogs of northern Scotland. The body was discovered by a pair of hikers on a cross-country trek. The person uncovered was approximately five feet six inches tall and appears to have been about 16 years old at the time of death. From the clothing still attached to the body, it is estimated that the teenager lived in the middle to late portion of the tenth century AD.

The body has been amazingly well preserved. In fact, Dr. Pure Research, one of the scientists in charge of the investigating team, said, "It would not surprise me a lot if this young person's eyes just opened right up and we were given a big smile of thanks."

Assume that the impossible did happen, that the teenager from the tenth century did suddenly awaken. If it did happen, someone would have to help the young person adapt to the changes in society and the world in the last ten centuries.

You have been selected to write a letter to this teenager explaining some of the environmental changes that have occurred since the tenth century. You may write your letter to either a 16-year-old girl or boy.

1. Form groups as instructed by your teacher.
2. Select one of the areas from the list below to write your letter about. No more than one member of each group may write about any one area.
   air quality
   water quality
   energy sources
   amount of forests
   types of materials available for building/clothing
   methods of transportation or communication
3. Put each group member's name on a piece of paper and after the name, list the area he or she selected.

(continued)

Scribe's Name ______________________ Date ______________

4. Don't Get Bogged Down on This *(continued)*

4. Do some research on England/Scotland/Europe in the tenth century and the world of today.

5. Write your letter. Be sure to include the following topics:

    - what your area was like when the teenager was alive (start with this to remind the poor dead person of what his or her life was like—after all, the person has not been alive for one thousand years!)
    - what your area is like now (what changes have occurred)
    - reasons (at least three) for the changes
    - your opinion of the changes (good or bad)

Use your creative ability. Write a rough draft in pencil and staple it to the back of your final draft. Your final draft should be written in ink or typewritten, using as many pages as necessary. Treat this assignment like an English writing assignment. Proofread your final draft for errors in spelling and grammar.

Scribe's Name ______________________ Date ______________

4. Don't Get Bogged Down on This *(continued)*

In-Class Assignment

1. Get back into the groups formed when you picked your areas to research.

2. Listen carefully while one person reads his or her letter out loud to the group.

3. Whenever the reader mentions the topic required in his or her letter, stop the reader and mark that place on the letter with one of the following codes:
   A (for ancient)—what your area was like when the teenager was alive
   N (for now)—what your area is like now
   R-1, **R-2**, **R-3** (for reasons)— at least three reasons for the changes
   O (for opinion)—your opinion of the changes

4. After each person finishes reading aloud, take about five minutes to write your reaction to his or her letter. Be sure to include one good thing in or about the letter, and one improvement you might make in the letter, and say whether you agree or disagree with the writer's opinion.

5. Give your reaction paper to the writer of the letter.

6. Continue to listen to, stop, and react to each member of the group.

7. Staple all of the letters underneath a sheet with a list of all of the names and areas of research from the first day, and turn them in to your teacher.

Scribe's Name ______________________________ Date ____________________

5. Science in (and as) History

There has never been a time in history when there have not been significant individuals who have contributed to the advancement of science. Your task is to:

1. Select a specific prominent scientist and do some research on that individual's life and achievements.
2. Do some additional research on the time period in which your scientist lived.
3. Determine how the scientist fit into the time in which he or she lived.
4. Explain how that particular time might have influenced the individual's "discovery" or scientific advancement.

Be sure to include in your paper all of the information asked for in 1–4 above. Your paper should have at least two paragraphs on each of the four numbered parts of this assignment. Your final draft should be written in ink or typewritten, using as many pages as necessary, but three pages of double-spaced typing is a good goal. Treat this assignment like an English writing assignment. Proofread your final draft for spelling and grammar errors.

Scribe's Name ______________________________ Date ____________________

6. My Life as a Fish

For this assignment, you will be asked to complete three parts in order as you produce a story. Be sure to follow each step so you will have a chance at receiving full credit on this assignment.

Part 1: Write *short* answers to each of the following questions.

1. How does a fish swim?
   a. Which fin is most important?
   b. In which direction does that fin move while the fish is swimming?
2. How does a fish breathe?
   a. What part of the fish's body is used in breathing?
   b. Where does the oxygen for the fish to breathe come from?
3. What are three reasons that a fish is covered with mucus?

Part 2: Now rewrite your answers to the above questions as complete sentences on a fresh sheet of paper.

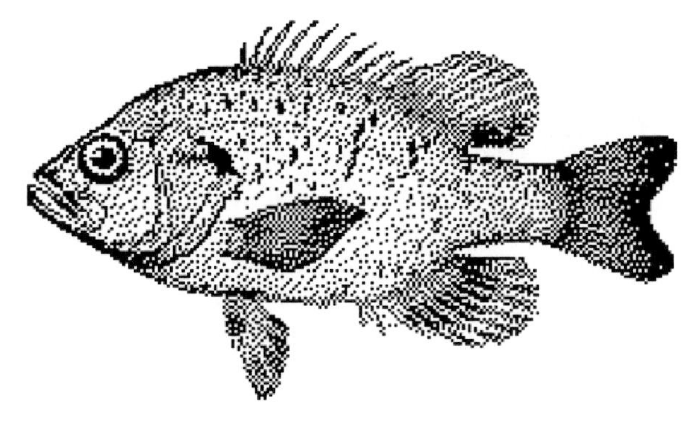

(continued)

Scribe's Name ______________________________ Date ____________________

6. My Life as a Fish *(continued)*

Part 3: Write a short, fishy, fictional story in pencil. You should pretend that you are a talking fish who is swimming in a pond. Here are the requirements for the story.

a. Describe how you are swimming in the pond.

b. Tell about the food you tried to eat.

c. Explain how it felt when the hook hidden in the food poked into your mouth.

d. Write down what you would say to the fisherman after he has you in his hand.

- Tell him why you need to go back into the water quickly.
- Tell him why he should not touch your slimy scales any more than he has to before letting you go.

Scribe's Name ______________________________ Date ____________________

7. Rain, Rain, Go Away!

San Diego, California. A fine place to visit. Lots of beaches. The San Diego Zoo and Wild Animal Park. Sea World. Over 300 days of sunshine annually. But San Diego does not have everything. Take water, for example.

The vast majority of San Diego's drinking water comes from outside of San Diego County. Water comes via aqueduct from the Colorado River and Northern California. Of course, some of the water used by San Diegans does fall from the sky. Just how much is shown below in the data table, which lists San Diego's annual rainfall totals for nearly 100 years.

Before you look at that data table, however, answer the following questions.

1. What is the average amount of rainfall you would expect a city like San Diego to get in a year? Why did you estimate that amount?

2. In what climate type or **biome** would you expect San Diego to be located?

Rainfall Data

Year	Annual Rainfall	Year	Annual Rainfall	Year	Annual Rainfall
——	——	1930–31	10.78	1970–71	8.00
1891–92	8.70	1931–32	13.18	1971–72	5.69
1892–93	9.26	1932–33	10.63	1972–73	11.43
1893–94	4.97	1933–34	4.26	1973–74	6.60
1894–95	11.90	1934–35	15.10	1974–75	10.65
1895–96	6.21	1935–36	8.39	1975–76	9.11
1896–97	11.78	1936–37	15.93	1976–77	8.08
1897–98	4.99	1937–38	9.72	1977–78	18.71
1898–99	5.24	1938–39	9.79	1978–79	15.55
1899–1900	5.97	1939–40	11.30	1979–80	15.72

(continued)

Scribe's Name ______________________________ Date ________________

7. Rain, Rain, Go Away! *(continued)*

Rainfall Data *(continued)*

Year	Annual Rainfall	Year	Annual Rainfall	Year	Annual Rainfall
1900–01	10.45	1940–41	24.74	1980–81	8.10
1901–02	6.17	1941–42	13.05	1981–82	11.50
1902–03	11.76	1942–43	11.10	1982–83	18.26
1903–04	4.40	1943–44	14.47	1983–84	5.73
1904–05	14.32	1944–45	11.04	1984–85	9.65
1905–06	14.68	1945–46	9.34	1985–86	14.10
1906–07	10.62	1946–47	6.33	1986–87	9.61
1907–08	8.55	1947–48	6.83	1987–88	13.18
1908–09	10.23	1948–49	10.42	1988–89	5.65
1909–10	9.79	1949–50	8.55	1989–90	7.84
1910–11	11.99	1950–51	5.92	1990–91	11.73
1911–12	10.75	1951–52	18.16		
1912–13	5.97	1952–53	6.54		
1913–14	9.83	1953–54	9.13		
1914–15	14.41	1954–55	7.21		
1915–16	12.55	1955–56	4.52		
1916–17	10.13	1956–57	8.89		
1917–18	8.04	1957–58	13.90		
1918–19	8.74	1958–59	5.28		
1919–20	8.91	1959–60	7.45		
1920–21	7.08	1960–61	3.46		
1921–22	18.65	1961–62	9.63		
1922–23	6.36	1962–63	3.98		
1923–24	5.66	1963–64	7.05		
1924–25	5.81	1964–65	8.50		
1925–26	15.66	1965–66	15.07		
1926–27	14.74	1966–67	10.63		
1927–28	8.71	1967–68	7.96		
1928–29	7.10	1968–69	11.60		
1929–30	10.73	1969–70	6.34		

After you have taken some time to look at the data table, answer the remaining questions.

3. How much rain fell in San Diego in the year you were born?

4. How many years did San Diego have annual rainfall in the following ranges?

(continued)

Scribe's Name ______________________ Date ______________

7. Rain, Rain, Go Away! *(continued)*

A. a. 0.00–3.00 inches
   b. 3.01–6.00
   c. 6.01–9.00
   d. 9.01–12.00
   e. 12.01–15.00
   f. 15.01+

B. Which of the range(s) above had the most years?

C. Make a **histogram** (like a bar graph) of the number of years in each range vs. the rainfall ranges.

5. What year(s) had the highest rainfall total? What was that amount?

6. What was the average annual rainfall for San Diego during this 100-year period?

7. Approximate annual rainfall totals of various world biomes are given in the list below. Unfortunately, the totals are given in centimeters, not inches, and you will have to convert the San Diego 100-year average into centimeters (1 inch = 2.54 cm). Based on rainfall totals only, to which of the biomes does San Diego belong?

   Deciduous Forest: 80–100 cm
   Tropical Rain Forest: up to 500 cm
   Desert: 0–25 cm
   Grasslands: 26–75 cm
   Coniferous Forest: 100–150 cm

8. There are places in the world, tropical rain forests, for example, that receive over 200 inches of rainfall annually. How many years of tropical rain forest rain would it take to reach San Diego's 100-year total?

9. Make a bar graph comparing the three decades 1910–1920, 1950–1960, and 1980–1990. In how many years did the decade of the 1950's have the heaviest rainfall of the three?

10. What five-year period had highest cumulative amount of rainfall? What was the five-year total? How much above the 100-year average was the average for this five-year period?

11. Which five-year period had the lowest cumulative amount of rainfall? What was the five-year total? How much below the 100-year average was the average for this five-year period?

12. Ask your teacher for the last year's annual rainfall for your city or town. How many years of average San Diego rainfall would it take to match your last year's amount?

Scribe's Name ______________________________ Date ____________________

8. Fluffy as a Cloud

DAY 1

Air moves mainly due to temperature differences; warm air rises and cold air sinks. As warm air rises, however, it begins to cool off. Cool air can hold less water vapor than warm air, so as rising air cools, it gets closer and closer to the point where it cannot hold any more water—the air's **saturation point**. When air cannot hold any more water, water droplets start to **condense**, or group together, and clouds form.

When air is saturated, its relative humidity is 100%. The temperature of saturated air is called the **dew point**. Clouds are made up of water droplets condensing and forming around a piece of dust, or some other small particle. Water droplets will not condense, and clouds cannot form, until the air temperature has reached the dew point (once again, when the air has 100% relative humidity and is saturated).

It is pretty easy to get a good idea of the dew point of the air at any time. The amount of water vapor in the air changes from day to day, even hour to hour. (A summer morning may feel pleasant and dry, while the afternoon is muggy and uncomfortable.) The dew point can vary from day to day or hour to hour, too. Water droplets collect on the outside of a glass of ice water because the air surrounding the glass has been cooled and has reached its dew point. You can use this simple observation to find out where clouds should be forming in the sky above you.

Procedures

1. Take a shiny can (no label) and half fill it with water, then add two or three large ice cubes. Stir the ice water gently with a thermometer until you first see condensation (water droplets) appearing on the outside of the can. Answer question 1.

2. Working quickly, remove the ice from the can and discard it in the sink. Dry the outside of the can with the paper towel and continue stirring the water gently. Check the temperature of the water; if condensation appears again, dry the can and continue stirring. You want to find the temperature of the water at which condensation no longer appears on the can. It should be only a few degrees different (higher) from the temperature at which condensation first appeared. Follow your teacher's instructions for emptying and storing the can. Answer question 2.

(continued)

Scribe's Name ______________________________ Date ____________________

8. Fluffy as a Cloud *(continued)*

3. Add your two recorded temperatures together, and divide the answer by two. This average temperature will be considered the dew point of the air around you right now. Answer question 3.

4. Answer questions 4–7.

Questions

1. What is the water temperature when condensation first appears?

2. What is the water temperature when condensation no longer appears?

3. What is your dew point temperature? (Show your calculations.)

4. What do you know about the relative humidity of the air at the dew point?

5. What is a term used to describe air and the water it holds at its dew point?

6. If the air temperature is higher than the dew point, is the relative humidity more or less than 100%?

7. If the air temperature dropped even lower than the dew point, what would you expect to happen?

Scribe's Name ______________________________ Date ____________________

8. Fluffy as a Cloud *(continued)*

DAY 2

You already know that warm air rises, and that it cools off as it rises. What might be surprising is that moist air cools off at a fairly even rate. For each 100 meters the air rises, its temperature will drop .6 degree Centigrade (or, for every 1000 feet it rises, its temperature will drop 5.5 degrees Fahrenheit). When air has risen high enough in the sky to cool off and reach its dew point, clouds will form.

Just from looking up in the sky you know that clouds have many different shapes (besides the fact that some look like trees and others look like bunnies). Some clouds are puffy and round, others are thin streaks, some look like fog. These different shapes or types of clouds usually form at different levels above the ground, at different altitudes. In fact, cloud types are named according to their altitude and their shape. For the next part of the activity you will need some reference pictures of the different types of cloud formations.

Procedures

1. Take your reference pictures outside and compare them to the clouds in the sky above you. Find the best picture match. Answer questions 1 and 2.

2. Measure today's air temperature at ground level. Answer question 3.

3. Using the same procedure you used on Day 1, determine the dew point for today's air. Answer questions 4 and 5.

4. Using your air temperature reading, your determination of dew point, and the figures given above for the change in air temperature with altitude, calculate the altitude needed to reach the dew point (the expected altitude of the clouds in the sky). See the example below.

   Example: Air Temperature = 25°C Calculated Dew Point = 5°C

   25 – 5 = 20°C difference

   20 degrees C difference / .6 degrees C change × 100 m = 20 / .6 × 100 m =

   33 × 100 m = 3300 m = 3.3 km (expected altitude of clouds in the sky now)

5. Answer questions 6, 7, and 8.

(continued)

Scribe's Name ______________________ Date ______________

8. Fluffy as a Cloud *(continued)*

Questions

1. What type of clouds are in the sky?
2. What is their approximate altitude? (Use your reference material to find this answer.)
3. What is the air temperature (be sure to tell whether it's in degrees Fahrenheit or Centigrade).
4. What is the dew point?
5. Make a general statement relating altitude and air temperature.
6. Based on your calculations, at what altitude do you expect clouds to form right now?
7. What type of clouds should form at this altitude?
8. Did your observations and your calculations agree? That is, were the clouds you observed in the sky found at pretty much the same altitude you calculated for the dew point?

Scribe's Name ________________________________ Date ____________________

9. My Life as a Cloud

People "see" many different things when they look up at clouds in the sky—animals, famous people, everyday objects—but whether a cloud looks like a horse or a teapot, it is formed in the same way.

Use a book and your knowledge of clouds and how they form to answer the following questions.

1. Why do water droplets join together to form a cloud?
2. How do water droplets join together?
3. Name and describe the three basic kinds of cloud formations.
4. At what approximate altitude does each kind of cloud form?
5. Why might a cloud type change and become another kind of cloud?
6. What is a cloud at ground level called?

Now, imagine you are a droplet of water vapor. Use your creative ability and describe in a story how you would help form a cloud and the different kinds of cloud you could become. Be sure to include in your story all of the information you used to answer the questions above. Your final draft should be written in ink or typewritten using as many pages as necessary. Treat this assignment like an English writing assignment. Proofread your final draft for spelling or grammar errors. Staple the answers to the questions above to the back of your final draft.

Scribe's Name ______________________________ Date ____________________

10. Let's Get Mineralized

The long tongue of the herbivorous reptile wrapped its rough surface around a small clump of marsh grass. With a delicate tug and a quick twist of the reptile's neck, the grass was uprooted and pulled into the animal's mouth. The beak-like oral opening moved up and down and side to side simultaneously, allowing molars to mash the blades of grass into a pulp before the slobbery green mass was swallowed.

A sharp snort alerted the medium-sized dinosaur and halted its breakfast in mid-swallow. The horselike head that extended above narrow shoulders swiveled toward the sound, the large nostrils flaring in a combination of fear and anger. Standing at the edge of swamp was the animal who had snorted its challenge.

From the edge of the swamp, the young **Allosaurus** *could see what it hoped would become its breakfast. When the* **Camptosaurus's** *nostrils flared, so did those of the carnivore. But the nostril flaring of the carnivore had no fear associated with it. A second snort exploded from the Allosaurus. It sprang toward its prey.*

The carnivore's feet plunged through the thin layer of water that covered the mud of the swamp. Immediately, the thick, black goo oozed between the toes of the attacker, slowing its progress. As the Allosaurus began its attack, the Camptosaurus began a graceful swing of its tapered tail.

A sucking, splashing sound announced every approaching footstep of the hungry carnivore. But, each footstep was slightly slower than the one before it. In spite of the tremendous strength of the animal and its determination to succeed in its attack, the mud of the swamp showed no favoritism. Just as it slowed the movement of the herbivores who ate its grasses, the thick swamp mud slowed the advance of the predator as it moved toward its prey.

By the time the carnivore was within reach of its hoped-for breakfast, it was moving at about half of its normal attack speed. It was also slightly off balance as it fought against the tug of the gooey mud. The slowed attack speed and unbalanced movement of the Allosaurus and the heavy, swinging tail of the Camptosaurus produced a lethal combination for the attacker.

The herbivore's tail made contact once with the skull of the carnivore, but one touch was all the leadlike tail needed. The tip of the Camptosaurus's tail smacked across the eye socket of the charging carnivore. At that instant, the Allosaurus lurched toward the oncoming missile as it fought to maintain an attack posture. The effect of the lurch toward the swinging tail was the same as that of a fighter leaning into a punch; it multiplied the incredible force of the tail's hit and knocked out the Allosaurus.

(continued)

Scribe's Name ________________________________ Date ____________________

10. Let's Get Mineralized *(continued)*

For a brief instant the attacking dinosaur swayed, still upright. But unconsciousness moved swiftly from the brain down the neck and back of the carnivore. A tremendous splash of water and slimy goo announced its facedown landing in the swamp. The automatic breathing mechanism continued to inhale and exhale. All that was drawn into the nose and lungs of the Allosaurus was muddy liquid. Without the vast quantities of oxygen needed to sustain the life of the mighty hunter, the body shuddered and jerked once in a death spasm. Then it lay still.

The Camptosaurus was not aware that its attacker was dead. It only realized that it had gained some time to begin an escape. Sucking sounds like those heard moments before from the predator now announced the retreat of the prey.

At that moment, the mist that had been falling turned into a downpour. Runoff from the storm was already beginning its silt-laden rush down the nearby mountain slopes. The swamp would soon become a lake again.

* * * * *

The scenario above happened millions of times during the Jurassic period of geologic time. It was a time of mighty hunters like *Tyrannosaurus* rex and enormous herbivores like *Diplodocus.* From the smallest dinosaur to the largest, life was a struggle. And, even though the dinosaurs were the dominant land animals for over 100 million years, nearly all of the evidence of their existence left today is in the form of **fossils**.

Fossils are the mineralized remains or traces of an organism. The process of fossilization takes a long time and just the right conditions. Even if conditions are perfect for fossil formation, however, not all parts of an organism can be fossilized. Bones make nice fossils. Muscles do not.

Your task in this exercise is to explain how a fossil is formed from the fossil's point of view. Pretend you are the *Allosaurus* in the story above. What would happen to your body as it lay in the fast-filling swamp? How long would it take to become a fossil? What else might you become besides a fossil?

Be creative as you describe the process of fossilization. The explanation should be fairly detailed, but not boring. Be sure to answer the three questions in the paragraph directly above this one. Finally, describe how and when you are discovered and how it feels to be famous.

Write a rough draft in pencil and staple it to the back of your final draft. Your final draft should be written in ink or typewritten using as many pages as necessary. Treat this assignment like an English writing assignment. Proofread your final draft to be sure that you have no errors in spelling or grammar.

Scribe's Name ______________________ Date ______________

10. Let's Get Mineralized *(continued)*

In-Class Individual Assignment

1. On a separate sheet of paper, answer the following questions with information you remember from your story.
2. Questions:
   a. What is the first step in the formation of a fossil?
   b. How long does it take for a fossil to form (on the average)?
   c. What parts of the dead dinosaur usually become fossils?
   d. If you had not become a fossil, what other material might you have become?
   e. Where in the world were you most likely discovered?
   f. If you had been a female dinosaur who had just laid some eggs before dying, would the eggs fossilize? Why or why not?
3. Listen for further instructions.

Scribe's Name ____ Date __________________________

10. Let's Get Mineralized *(continued)*

In-Class Group Assignment

1. Pick up your papers from your teacher and form groups as instructed by your teacher.
2. Listen carefully while one person reads his or her story out loud.
3. Stop the reader as he or she answers each of the three required questions from the instruction page. Mark a star (*) on the paper at each place where part of the assignment is explained.
4. After each person finishes reading aloud, take about five minutes to write your reaction to his or her story. Be sure to include one good thing in or about the story and one improvement you might make in the story.
5. Give your reaction paper to the writer of the story to read.
6. Continue to listen, stop, and react to each member of the group.
7. Select one member of the group to be a scribe.
8. Write each group member's name on a sheet of notebook paper or use the consensus sheet your teacher gives you.
9. Write a group consensus answer to each of the questions.
10. Staple the stories underneath the consensus answers in the order that the group members' names are listed on the answer sheet.
11. Give the packet to your teacher.

Scribe's Name ________________________________ Date ____________________

11. My Fossil's Older Than Your Fossil

Fossils are the evidence of organisms that existed millions of years ago. Fleshy parts of animals usually do not fossilize well, but bones and teeth do a fine job of fossilization. Some times, of course, an entire organism is fossilized, but this is very rare. More common whole specimens are those in which an organism has been trapped in something like **amber** (a fossil resin). We can also see what an organism looked like if it left an imprint in mud that hardened and later filled with some substance that assumed the form of the dead organism.

In the exercise that follows, you will have a chance to see how well you can judge which fossil came before which. You will base your judgment on pictures of an artist's reconstructions of a variety of fossilized remains found in the eroded bank of a river.

Part 1

1. Get into groups as instructed by your teacher.
2. Cut out the pictures of the organisms and arrange them in what you think would be the most logical sequence of development or progression. Assume that one fossil was the ancestor of all the rest and that other organisms were descendants of *specific* other organisms. Use the structures (antennae, tails, etc.) shown on the organisms in the pictures to help you in your decision-making. Also use the following clues:

   Clue #1: From the original organism (*Tortugis ancientis*), three distinct families branched off and founded their own lines of development.

   Clue #2: New organisms do not have to have formed in different lines at the same time. They *might* have, but they do not *have* to have.

   Clue #3: Each of the original three lines *could* have more than one branch.
3. After you have arranged your organisms in your "best guess" order, write down the names of the organisms in whatever pattern you decided was the best (like a family tree). Save the pictures!
4. Take the paper with the written family tree to your teacher when your group is done.

Good Luck!

(continued)

Scribe's Name ______________________________ Date ____________________

11. My Fossil's Older Than Your Fossil *(continued)*

Illustrations of Tortugis *Species*

Tortugis ancientis

Tortugis predatoris

Tortugis slidereni

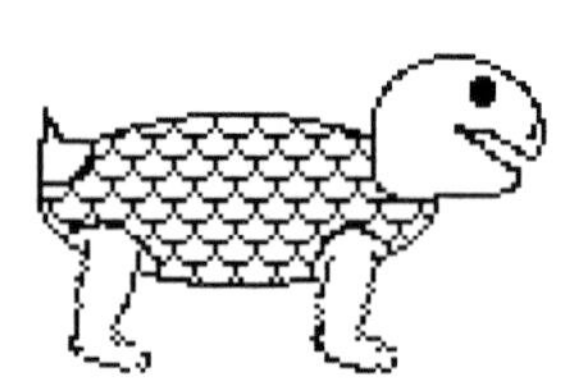

Tortugis scalifiedis

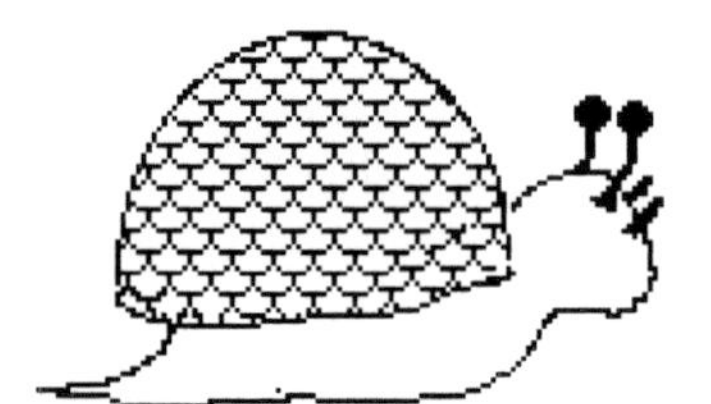

Tortugis molluskus

Tortugis proboscis

Tortugis porkii

Tortugis spinalis

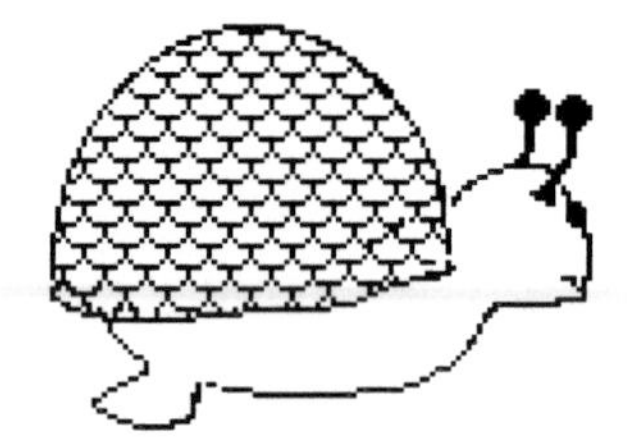

Tortugis apodis

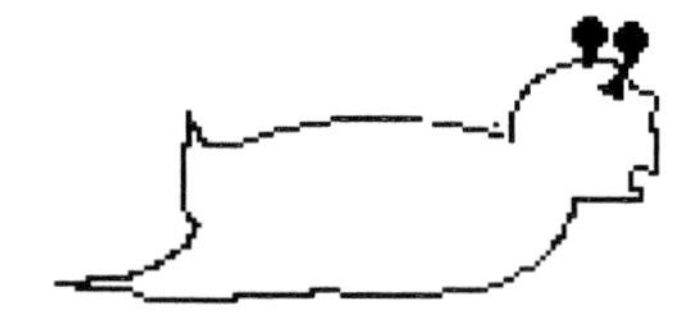

Tortugis slugenii

Tortugis tortugis

Scribe's Name ____________________________ Date ____________________

11. My Fossil's Older Than Your Fossil *(continued)*

Part 2

Congratulations! You completed the first part of this assignment. Now you will get a chance to see where the fossils were found and consider modifying any of your answers before you turn them in.

Below is a diagram of the river bank strata where the *Tortugis* fossils were discovered. Below the diagram is a list of **strata layers** and fossil names. Each fossil name appears after the number of the layer in which it was discovered. You may assume that the layer labeled with a name is the lowest layer in which that particular type of fossil was found.

Surface

6

5

4

3

2

1

Stream

Stratum number	Fossil name
1	*Tortugis ancientis*
2	*Tortugis scalifiedis*
3	*Tortugis slidereni, Tortugis tortugis*
4	*Tortugis slugenii, Tortugis proboscis*
5	*Tortugis spinalis, Tortugis apodis, Tortugis porkii*
6	*Tortugis molluskus, Tortugis predatoris*

Redraw your family tree as needed to match this new information.

(continued)

Scribe's Name ________________________ Date ________________

11. My Fossil's Older Than Your Fossil *(continued)*

Important note: Just because the name of one of the *Tortugis* organisms appears directly above another name on the list, it does not mean that those two organisms are directly related.

Now answer the following questions as assigned by your teacher.

1. Which were the oldest fossils after *Tortugis ancientis*? How did you decide that?
2. In which layers were the most *complex* fossils located?
3. Could there be species of *Tortugis* other than those that have been found? Explain why or why not.
4. Look carefully at each of the *Tortugis* specimens. Write down what you think each one ate and where it lived. Explain why you made the decisions you did.
5. a. Choose any two consecutive steps in the family tree you ended up with. Make a drawing of one other type of *Tortugis* that would be like a missing link between the two that are on your family tree.

   b. Name your creature *Tortugis* ________________. Choose a species name that would be as appropriate for your organism as the species names for the known specimens.

Scribe's Name ______________________________ Date ____________________

12. There'll Be a Hot Time in the Old Town Tonight!

Uleokoi stood in the door of his hut. He held his infant daughter in his arms and gazed up the slope of the towering volcano that dominated his island home. Oahoa, goddess of the volcano, was unhappy. She had been unhappy for several days, but tonight it was obvious that she wanted some new kind of sacrifice.

It should have been completely dark. The moon god, Malakahai, was away on his monthly journey to regain his light, so there was no moon. Only millions of stars should have cast their pitiful radiance down on the earth. Instead Uleokoi stared through the red glow of the volcano that reflected from the clouds of smoke and ash that had blocked out even the sun for the past two days.

The islander shifted the weight of the little girl in his arms. Gently he brushed the ash from her jet-black curls. The ground rumbled as the mountain belched out another cloud of angry gases. He felt a hand on his shoulder and knew it was Loloi, his wife. Instinctively he put his arm around her. Though the tropical night was warm, she shivered. He pulled her closer to his side.

He looked around at his village. About a dozen grass huts were randomly spaced among the stately palm trees that produced the coconuts the villagers used as one of their staples. The carved stone god, Ahoa, protector of the fishermen, stood in silent watch over the dugout canoes that were lined up in the sand at the ocean's edge. All of the buildings he observed had gently rounded edges. Except one.

The coral-block dispensary-home of the missionary stood at the edge of the village. The square corners and coral-block walls of the small rectangular building were a dramatic contrast to the circular grass huts. Only the palm frond roofs were the same.

Whether it was the shock of sound waves or the shaking of the ground from the massive volcanic eruption that threw the family to the ground, Uleokoi never knew. His daughter began screaming in fear. His wife's fingers closed around the shell armband he wore, pushing the sharp edges of the shells into his arm. More screams echoed from the other huts as the shaking continued.

Uleokoi raised himself to his knees and once again turned to look at the mighty mountain, but he could not see it. A fast-moving cloud of ash and gas raced down the mountainside, obscuring it from his sight. It looked like the mountain was attacking the village.

In spite of all his courage, Uleokoi opened his mouth and screamed. His cry and those of his friends and relatives were cut short as the gases from the eruption burned the lungs of the villagers, suffocating them all in seconds.

(continued)

Scribe's Name ______________________________ Date ____________________

12. There'll Be a Hot Time in the Old Town Tonight! *(continued)*

The white-hot ash fell like no snowfall ever had. In an hour the entire village was covered in nearly three feet of powder. By the next morning the lava flow from the volcano reached the sea, adding billows of steam to the smoky clouds that covered the island.

* * * * *

"Dr. Branton, look!" The call to the leader of the archeological expedition was filled with excitement.

"What is it Dave?" the doctor asked as she moved toward her student assistant. She had been doing this kind of work on the Pacific islands ever since her graduation in 2068 and had yet to make a significant discovery. She was about to conclude that all of the major archeological finds had already been made. She had learned to keep her excitement under control.

"I've chipped through this lava crust," the college student answered breathlessly. "And, look, here is some solidified ash," he continued tapping the gray-white rock. "But here is a hole. And it is a big one!"

"That could be important," Dr. Branton admitted. This did have some promise. She knelt down, turned on her flashlight, and peered under the amazingly thin crust of lava into an irregularly-shaped depression. She reached in and felt around.

"I think we found where a body once lay," she concluded after several seconds of probing. The forced calm of her first comment finally gave way to her true feelings, and she nearly shouted, "Good work, Dave! This must have been a village once!"

* * * * *

Your assignment in this exercise is to write a description of what Dr. Branton and her team uncovered on the island. Write your answer like a journal or diary that one of the team members might have written during the dig.

Be sure to address the following points in your journal:

1. What remains (objects, implements, etc.) from the village would have survived beneath the ash? (You can assume that the lava flowed *over* the ash layer. You can also assume that Ahoa was knocked face-down into the sand by the earthquake before the ash arrived.)

2. Why would those items have survived?

(continued)

Scribe's Name ______________________________ Date ____________________

12. There'll Be a Hot Time in the Old Town Tonight! ***(continued)***

3. What would *not* have survived, and why?

4. Pretend you are an archeologist and draw a diagram of the village based on your journal entries and observations while excavating the site. Mark on the diagram what objects and implements you found and where you found them.

Use your creative ability. Write a rough draft in pencil and staple it to the back of your final draft. Your final draft should be written in ink or typewritten using as many pages as necessary. Treat this assignment like an English writing assignment. Proofread your final draft to be sure that you have no errors in spelling or grammar.

Scribe's Name ______________________________ Date ____________________

12. There'll Be a Hot Time in the Old Town Tonight! *(continued)*

In-Class Student Instructions

1. Pick up your papers from your teacher and form groups as instructed by your teacher.

2. Listen carefully while one person reads his or her story out loud to the group.

3. Stop the reader as he or she covers one of the required points in the assignment. Mark a star (*) on the paper at each place where part of the assignment is explained.

4. After each person finishes reading aloud, take about five minutes to write your reaction to each story. Be sure to include one good thing in or about the story and one improvement you might make in the story.

5. Give your reaction paper to the writer of the story.

6. Continue to listen, stop, and react to each member of the group.

7. Select one member of the group to be a scribe.

8. Write each group member's name on a sheet of notebook paper or use the consensus sheet your teacher supplies.

9. Write a group consensus answer to each of the questions.

10. Staple the stories underneath the consensus answers in the order that the group members' names are listed on the answer sheet.

11. Give the packet to your teacher.

Scribe's Name ______________________ Date ______________

13. "Time" for a Vacation Trip

When you get ready to take a vacation, you might visit a travel agent to help plan air or ground travel, to reserve a hotel room, or even to pick the ideal vacation spot. Because of this, the walls of travel agencies are covered with beautiful posters of exotic vacation locations.

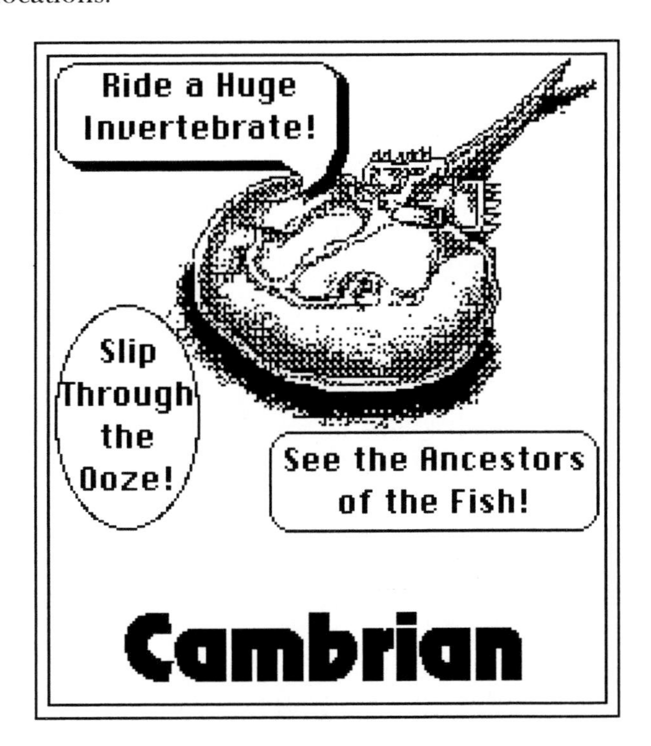

Your task in this exercise is to design a colorful travel poster for one of the geologic time periods in our earth's history. The poster should be no larger than 11 × 17 inches. It should have at least one large picture of a significant feature of the time period. There should be between three and five slogans on the poster. Each slogan should describe in words some aspect of the time period. One of the slogans could describe some make-believe amusement-parklike ride that might be found there. Each slogan should be inside some design, like the words of a cartoon character or the sound when something explodes in a comic strip. The name of the geologic period should be at the very top or bottom, or down the side of the poster in large letters.

Check with your teacher to see if you can choose any geologic period you want or if you have to sign up for a specific geologic period. Check also to see if there is a short written portion to this assignment.

Scribe's Name ______________________________ Date ____________________

14. Sounds Pretty Deep to Me

When people first began to wonder what the ocean bottom looked like, the best method available was simply called **sounding**. See Diagram 1 below.

A weight on a line was lowered from a ship until it hit the ocean bottom, then the length of the line was measured. This method was not particularly accurate because if the ship drifted, the line would be stretched out at an angle, making the ocean bottom appear deeper than it really was. Also, sometimes the line just was not long enough. Then too, mariners had to take a large number of separate readings to get a good idea of what the bottom looked like.

In this activity you will be drawing a profile of your personal section of the bottom of a mythical ocean. You will be given a set of **coordinates** similar to actual sounding data. You will plot these coordinate points on a piece of graph paper to produce a profile of your section of the ocean bottom.

Diagram 1. A ship, sounding

Procedure

1. Get into groups as instructed by your teacher.
2. Read through this entire procedure.
3. Assign one group member to be the plotter and the other to be the coordinate reader.
4. Get a piece of graph paper and a set of coordinates from your teacher. The scale of your coordinate set is 1" = 100m.

(continued)

Scribe's Name ______________________________ Date ____________________

14. Sounds Pretty Deep to Me *(continued)*

5. Write down the letter of the coordinate set and the names of each group member on your graph paper.

6. Label the *x*-axis "Distance from Starting Point." Label the *y*-axis "Depth Below Surface."

7. Mark off both the *x*- and *y*-axis in the scale given in step 4.
   Note: You will be plotting points measured from the surface of the ocean downward. The positive direction on the *y*-axis is down. Zero will be the ocean surface level, the *x*-axis should be drawn near to the top edge of your graph paper.

8. The coordinate reader reads the first set of coordinates aloud.

9. The plotter plots the point on the graph paper indicated by the first set of coordinates.

10. Repeat steps 8 and 9 for all of the sets of coordinate points.

11. Connect the points plotted.

12. Take your completed profile and compare it with the complete ocean floor profile your teacher has.

13. Answer the questions below for your profile on a separate sheet of smooth-edged notebook paper. Staple the answers to the questions to your profile.

Questions to be answered for your profile

1. What number on the complete profile matched the profile you drew?

2. Why did the profile you drew not look exactly like the section of the completed profile it represented?

3. What could you change about the procedure to get a more accurate profile of the actual sea floor?

4. What could you change about the data collection to get a more accurate profile of the actual sea floor?

(continued)

Scribe's Name ______________________ Date ______________

14. Sounds Pretty Deep to Me *(continued)*

5. What would you expect to happen to your profile if:

   a. there were twice as many coordinate sets as you received?

   b. there were one-half as many coordinate sets as you received?

6. What was the minimum depth on your profile?

7. What was the greatest depth on your profile?

8. Check with your teacher to see if you are supposed to use colored pencils to create a realistic-looking environment in your section of the ocean bottom.

Scribe's Name ______________________________ Date ____________________

14. Sounds Pretty Deep to Me *(continued)*

Coordinate Listing for Ocean Bottom—Section A

x-axis	*y*-axis
0	2.0
.25	2.0
.50	1.7
.75	1.5
1.00	2.0
1.25	1.9
1.50	1.8
1.75	1.2
2.00	0.9
2.25	0.8
2.50	0.8
2.75	1.3
3.00	1.8
3.25	2.5
3.50	2.8
3.75	2.6

x-axis	*y*-axis
4.00	2.0
4.25	1.7
4.50	1.8
4.75	1.8
5.00	2.0
5.25	2.0
5.50	2.2
5.75	2.0
6.00	2.0
6.25	2.3
6.50	2.3
6.75	2.0
7.00	2.0
7.25	2.2
7.50	2.8
7.75	2.0
8.0	2.0

Coordinate Listing for Ocean Bottom—Section B

x-axis	*y*-axis
0	2.0
.25	1.7
.50	1.5
.75	1.5
1.00	1.5
1.25	1.8
1.50	2.0
1.75	2.5
2.00	2.6
2.25	2.5
2.50	2.0
2.75	1.9
3.00	1.5
3.25	1.5
3.50	1.5
3.75	1.8

x-axis	*y*-axis
4.00	2.0
4.25	2.6
4.50	2.6
4.75	3.0
5.00	3.0
5.25	3.0
5.50	2.5
5.75	2.5
6.00	2.0
6.25	1.8
6.50	1.7
6.75	2.0
7.00	2.3
7.25	2.3
7.50	2.0
7.75	2.0
8.00	2.0

Scribe's Name ______________________________ Date ____________________

14. Sounds Pretty Deep to Me *(continued)*

Coordinate Listing for Ocean Bottom—Section C

x-axis	*y*-axis
0	2.0
.25	2.0
.50	1.8
.75	2.0
1.00	2.0
1.25	2.0
1.50	1.8
1.75	2.0
2.00	2.0
2.25	2.5
2.50	2.0
2.75	2.0
3.00	1.8
3.25	1.8
3.50	2.8
3.75	2.8

x-axis	*y*-axis
4.00	2.5
4.25	2.8
4.50	2.0
4.75	2.0
5.00	1.8
5.25	2.0
5.50	1.8
5.75	1.8
6.00	1.8
6.25	1.5
6.50	1.3
6.75	1.3
7.00	1.3
7.25	1.8
7.50	1.9
7.75	2.0
8.00	2.0

Coordinate Listing for Ocean Bottom—Section D

x-axis	*y*-axis
0	2.0
.25	1.8
.50	2.0
.75	2.3
1.00	2.5
1.25	2.3
1.50	2.3
1.75	2.5
2.00	2.3
2.25	2.0
2.50	1.8
2.75	1.5
3.00	1.5
3.25	1.8
3.50	2.0
3.75	2.6

x-axis	*y*-axis
4.00	2.8
4.25	3.0
4.50	2.9
4.75	2.8
5.00	2.5
5.25	1.4
5.50	1.3
5.75	1.3
6.00	1.6
6.25	2.5
6.50	2.3
6.75	2.3
7.00	2.3
7.25	2.6
7.50	2.2
7.75	2.0
8.00	2.0

Scribe's Name ______________________________ Date ____________________

14. Sounds Pretty Deep to Me *(continued)*

Coordinate Listing for Ocean Bottom—Section E

x-axis	*y*-axis
0	2.0
.25	2.0
.50	2.0
.75	1.8
1.00	1.8
1.25	1.9
1.50	1.8
1.75	1.3
2.00	1.3
2.25	1.3
2.50	1.7
2.75	1.9
3.00	2.0
3.25	2.0
3.50	2.1
3.75	2.1

x-axis	*y*-axis
4.00	2.0
4.25	2.0
4.50	2.0
4.75	2.3
5.00	2.7
5.25	2.9
5.50	3.0
5.75	2.9
6.00	3.0
6.25	2.8
6.50	2.8
6.75	2.4
7.00	2.1
7.25	2.1
7.50	2.0
7.75	2.0
8.00	2.0

Coordinate Listing for Ocean Bottom—Section F

x-axis	*y*-axis
0	2.0
.25	2.0
.50	1.6
.75	1.4
1.00	1.1
1.25	0.8
1.50	0.5
1.75	0.3
2.00	0.1
2.25	0.0
2.50	0.1
2.75	0.5
3.00	0.9
3.25	1.4
3.50	2.1
3.75	2.3

x-axis	*y*-axis
4.00	2.6
4.25	2.0
4.50	2.1
4.75	2.5
5.00	2.1
5.25	2.0
5.50	2.0
5.75	2.6
6.00	2.8
6.25	3.1
6.50	3.6
6.75	3.6
7.00	3.5
7.25	3.5
7.50	2.0
7.75	2.0
8.00	2.0

Scribe's Name ______________________ Date ______________

14. Sounds Pretty Deep to Me *(continued)*

Coordinate Listing for Ocean Bottom—Section G

x-axis	*y*-axis
0	2.0
.25	2.0
.50	1.8
.75	2.3
1.00	2.0
1.25	1.5
1.50	1.5
1.75	2.0
2.00	2.3
2.25	2.5
2.50	2.8
2.75	2.5
3.00	2.8
3.25	2.5
3.50	2.3
3.75	2.8

x-axis	*y*-axis
4.00	2.5
4.25	2.8
4.50	2.3
4.75	1.5
5.00	1.5
5.25	2.0
5.50	2.0
5.75	1.8
6.00	1.5
6.25	1.5
6.50	2.2
6.75	2.0
7.00	2.0
7.25	2.3
7.50	2.3
7.75	2.0
8.00	2.0

Coordinate Listing for Ocean Bottom—Section H

x-axis	*y*-axis
0	2.0
.25	1.9
.50	1.7
.75	1.5
1.00	1.5
1.25	1.5
1.50	2.0
1.75	2.0
2.00	2.0
2.25	2.0
2.50	2.2
2.75	2.5
3.00	2.5
3.25	3.0
3.50	2.9
3.75	2.7

x-axis	*y*-axis
4.00	2.7
4.25	3.0
4.50	2.9
4.75	2.2
5.00	2.0
5.25	1.8
5.50	1.8
5.75	1.7
6.00	1.5
6.25	1.5
6.50	1.3
6.75	1.3
7.00	1.0
7.25	0.9
7.50	2.0
7.75	2.0
8.00	2.0

Scribe's Name ______________________________ Date ____________________

14. Sounds Pretty Deep to Me *(continued)*

Coordinate Listing for Ocean Bottom—Section I

x-axis	*y*-axis
0	2.0
.25	1.7
.50	1.5
.75	1.3
1.00	1.0
1.25	0.8
1.50	0.5
1.75	0.7
2.00	2.1
2.25	2.6
2.50	2.6
2.75	2.8
3.00	3.3
3.25	3.5
3.50	3.5
3.75	3.5

x-axis	*y*-axis
4.00	3.6
4.25	3.6
4.50	3.5
4.75	3.5
5.00	3.2
5.25	3.0
5.50	2.8
5.75	2.5
6.00	2.4
6.25	2.0
6.50	1.8
6.75	1.9
7.00	2.2
7.25	2.2
7.50	2.0
7.75	2.0
8.00	2.0

Coordinate Listing for Ocean Bottom—Section J

x-axis	*y*-axis
0	2.0
.25	0.5
.50	0.3
.75	0.4
1.00	2.8
1.25	3.4
1.50	2.9
1.75	2.8
2.00	2.2
2.25	2.3
2.50	2.5
2.75	2.8
3.00	2.6
3.25	2.5
3.50	2.7
3.75	2.7

x-axis	*y*-axis
4.00	2.2
4.25	1.7
4.50	0.6
4.75	0.5
5.00	0.8
5.25	1.8
5.50	2.0
5.75	2.0
6.00	2.0
6.25	1.7
6.50	1.8
6.75	2.0
7.00	2.0
7.25	2.0
7.50	1.7
7.75	2.0
8.00	2.0

Scribe's Name ______________________ Date ______________

14. Sounds Pretty Deep to Me *(continued)*

Coordinate Listing for Ocean Bottom—Section K

x-axis	*y*-axis
0	2.0
.25	2.0
.50	2.0
.75	2.0
1.00	3.9
1.25	4.0
1.50	4.0
1.75	3.6
2.00	2.0
2.25	2.0
2.50	2.0
2.75	1.8
3.00	1.5
3.25	1.3
3.50	1.3
3.75	1.6

x-axis	*y*-axis
4.00	1.8
4.25	2.0
4.50	2.6
4.75	2.8
5.00	2.4
5.25	2.5
5.50	2.5
5.75	2.3
6.00	2.0
6.25	2.3
6.50	3.3
6.75	3.5
7.00	3.4
7.25	2.8
7.50	2.0
7.75	2.0
8.00	2.0

Coordinate Listing for Ocean Bottom—Section L

x-axis	*y*-axis
0	2.0
.25	2.0
.50	1.8
.75	2.0
1.00	1.3
1.25	2.0
1.50	1.8
1.75	2.0
2.00	2.3
2.25	2.5
2.50	2.3
2.75	2.0
3.00	1.7
3.25	1.5
3.50	1.75
3.75	2.0

x-axis	*y*-axis
4.00	2.3
4.25	2.6
4.50	2.8
4.75	2.3
5.00	2.0
5.25	1.8
5.50	1.7
5.75	1.3
6.00	1.5
6.25	1.8
6.50	2.8
6.75	2.8
7.00	3.1
7.25	3.1
7.50	2.8
7.75	2.0
8.00	2.0

Scribe's Name ______________________________ Date ____________________

14. Sounds Pretty Deep to Me *(continued)*

Coordinate Listing for Ocean Bottom—Section M

x-axis	*y*-axis
0	2.0
.25	2.0
.50	2.0
.75	2.4
1.00	3.0
1.25	3.4
1.50	3.7
1.75	3.8
2.00	3.8
2.25	3.8
2.50	3.4
2.75	3.1
3.00	2.5
3.25	2.5
3.50	2.5
3.75	2.8

x-axis	*y*-axis
4.00	3.1
4.25	3.3
4.50	3.3
4.75	2.8
5.00	2.5
5.25	2.3
5.50	2.0
5.75	1.8
6.00	1.5
6.25	1.4
6.50	1.3
6.75	1.3
7.00	1.5
7.25	1.9
7.50	2.0
7.75	2.0
8.00	2.0

Coordinate Listing for Ocean Bottom—Section N

x-axis	*y*-axis
0	2.0
.25	2.0
.50	1.8
.75	1.6
1.00	1.5
1.25	1.5
1.50	1.6
1.75	1.8
2.00	2.0
2.25	2.4
2.50	3.3
2.75	3.3
3.00	2.3
3.25	2.0
3.50	1.8
3.75	1.6

x-axis	*y*-axis
4.00	1.4
4.25	1.2
4.50	1.0
4.75	1.0
5.00	1.4
5.25	1.8
5.50	1.7
5.75	1.5
6.00	1.4
6.25	1.5
6.50	1.8
6.75	2.1
7.00	2.3
7.25	2.2
7.50	2.0
7.75	2.0
8.00	2.0

Scribe's Name ______________________ Date ______________

14. Sounds Pretty Deep to Me *(continued)*

Coordinate Listing for Ocean Bottom—Section O

x-axis	*y*-axis
0	2.0
.25	2.0
.50	1.4
.75	1.4
1.00	1.5
1.25	2.0
1.50	2.3
1.75	2.5
2.00	2.5
2.25	2.4
2.50	2.1
2.75	1.7
3.00	1.4
3.25	1.2
3.50	1.0
3.75	1.0

x-axis	*y*-axis
4.00	1.1
4.25	1.6
4.50	1.9
4.75	2.0
5.00	2.0
5.25	1.5
5.50	1.1
5.75	1.7
6.00	0.5
6.25	0.6
6.50	0.8
6.75	1.5
7.00	1.9
7.25	2.5
7.50	2.4
7.75	2.0
8.00	2.0

Coordinate Listing for Ocean Bottom—Section P

x-axis	*y*-axis
0	2.0
.25	2.0
.50	1.8
.75	1.8
1.00	1.8
1.25	2.1
1.50	2.5
1.75	2.8
2.00	2.8
2.25	3.1
2.50	3.1
2.75	2.8
3.00	2.4
3.25	2.3
3.50	1.8
3.75	1.0

x-axis	*y*-axis
4.00	1.0
4.25	2.1
4.50	2.5
4.75	2.6
5.00	3.0
5.25	3.0
5.50	3.0
5.75	3.0
6.00	3.0
6.25	2.3
6.50	1.3
6.75	1.3
7.00	1.8
7.25	2.3
7.50	2.3
7.75	2.0
8.00	2.0

Scribe's Name ________________________ Date ______________

14. Sounds Pretty Deep to Me *(continued)*

Coordinate Listing for Ocean Bottom—Section Q

x-axis	*y*-axis
0	2.0
.25	2.0
.50	2.3
.75	2.6
1.00	2.8
1.25	2.8
1.50	2.8
1.75	2.8
2.00	2.4
2.25	2.3
2.50	2.3
2.75	2.6
3.00	2.8
3.25	2.7
3.50	2.3
3.75	1.4

x-axis	*y*-axis
4.00	1.3
4.25	1.3
4.50	1.6
4.75	2.0
5.00	2.0
5.25	2.0
5.50	2.3
5.75	2.3
6.00	2.1
6.25	1.6
6.50	1.3
6.75	1.3
7.00	1.5
7.25	1.9
7.50	1.9
7.75	2.0
8.00	2.0

Coordinate Listing for Ocean Bottom—Section R

x-axis	*y*-axis
0	2.0
.25	2.0
.50	3.2
.75	3.3
1.00	3.2
1.25	3.3
1.50	3.0
1.75	3.0
2.00	3.4
2.25	2.8
2.50	2.8
2.75	3.5
3.00	3.1
3.25	2.4
3.50	2.0
3.75	2.0

x-axis	*y*-axis
4.00	2.5
4.25	3.1
4.50	3.5
4.75	3.5
5.00	3.2
5.25	2.9
5.50	2.4
5.75	1.9
6.00	1.3
6.25	1.0
6.50	1.0
6.75	1.6
7.00	2.2
7.25	2.5
7.50	2.3
7.75	2.0
8.00	2.0

Scribe's Name ______________________ Date ______________

15. Move Over, Gary Larson

The task in this exercise is to devise a cartoon on a topic in physical science. You may choose to do a four-frame cartoon (like *Peanuts* or *Cathy*) or a single-frame cartoon (like *The Far Side*). A third option is to do a political cartoon on some aspect of earth science. Check with your teacher to see if there is a list of acceptable topic areas for you to choose from before you begin your cartoon.

1. You may work on your cartoon by yourself or with a single partner.
2. The cartoon should be drawn, and colored, on a single sheet of unlined typing paper. All words and pictures should be dark or brightly colored.
3. Hopefully no explanation of your cartoon will be necessary.
4. The cartoon must be accurate in its portrayal of the earth science you use. For example, you cannot have igneous rock forming by sedimentation unless you correct that fallacy in the cartoon, etc.
5. You will have twenty-five minutes to complete this task. At the end of the time allowed for production, an opaque projector will show your cartoon to the entire class on the screen. Therefore, neatness counts.

Good Humor!

Scribe's Name ______________________________ Date ____________________

16. Rock On!

A geologist kneels on the rocky outcropping of a cliff wall high above a river. With skill and patience, the scientist chips away at the loose dirt surrounding a piece of rock. Once she has loosened the entire piece, she carefully removes it from its resting place of millions of years. She places it in a plastic bag to protect it while it is transported back to the laboratory, where it will be studied by a host of scientists and their assistants.

The sample of rock to be studied will fall into one of three general categories: **sedimentary**, **metamorphic**, or **igneous**. Each of these types of rock was formed by a different process. In this exercise you will be investigating one of those rock types and its formation by pretending to be it.

1. Get into groups as instructed by your teacher.
2. Get out one piece of smooth-edged notebook paper and put all group members' names on the top.
3. Decide which member of your group will become each type of rock. Unless you are told otherwise, only one group member can be each type of rock.
4. Write the type of rock that each member will become beside the name of that group member on the smooth-edged notebook paper from step 2.

Your task in this exercise is to explain how your type of rock is formed from the rock's point of view. Pretend you *are* the piece of rock in the introduction above. What forces helped to form you? How long did it take to make you a mature piece of your type of rock? What type of minerals and other material do you contain?

Be creative as you describe the process of rock formation. The explanation should be fairly detailed, but not boring. Be sure to answer the three questions in the paragraph directly above this one. Finally, describe your discovery and how it felt to be discovered. (Don't forget the hammer used to chip away the material surrounding you!)

Write a rough draft in pencil and staple it to the back of your final draft. Your final draft should be written in ink or typewritten using as many pages as necessary. Treat this assignment like an English writing assignment. Proofread your final draft to be sure that you have no errors in spelling or grammar.

Scribe's Name ______________________________ Date ____________________

16. Rock On! *(continued)*

In-Class Individual Assignment

1. On a separate sheet of paper, answer the questions below with information you remember from your story.

2. Questions:

   a. What is the first step in the formation of your type of rock?

   b. What physical force is the primary one responsible for the forming of your type of rock?

   c. What minerals are typically included in your type of rock?

   d. How did you get high above a river?

   e. Where in the world were you most likely discovered?

3. Listen for further instructions.

Scribe's Name ______________________________ Date ____________________

16. Rock On! *(continued)*

In-Class Group Assignment

1. Pick up your papers from your teacher and get back into the groups you were in when you decided which type of rock you would be.
2. During today's group work you will need your completed writing assignment, a piece of paper, and a pen or pencil.
3. You will have twenty minutes to teach each other how all of the rock types were formed. Be sure that each person describes the information asked for in the three questions in the writing assignment.
4. At the end of the twenty-minute teaching period there will be a ten-question quiz. There will be at least three questions about each of the types of rocks.
   a. If all members of your group get at least eight of the ten questions correct, each team member gets a bonus.
   b. If your group only has two people assigned, or if one of your group members is absent today, then the bonus will be awarded if the remaining two members each score at least seven correct.
5. If the entire class scores at least eight out of ten correct, there will be an additional bonus for each class member.
6. Before you take the quiz, staple the stories together in the order that the group members' names are listed on the cover sheet from the first day. Turn the packet in to your teacher.

Scribe's Name ________________________ Date ________________

17. Spend Some Time in the "O" Zone

"Don't use stuff if it comes in a spray can."

"Be sure to use sunscreen whenever you're in the sun."

"Will not harm the ozone layer."

You have probably heard or read words like those from friends or on commercials or product containers. All of those comments have something to do with the **ozone layer** that surrounds our planet. Ozone protects the earth from a variety of radiation, but most especially, from ultraviolet radiation.

There is significant concern among many scientists about depletion of the ozone layer. There is one hole in the layer that grows and shrinks over the Antarctic region. Some say that the hole grows bigger each year and shrinks to not quite as small when it shrinks. Newspapers report increases in skin irritation in people living in extremely southern South America.

Your task in this exercise is to design and draw a poster concerned with the depletion of the ozone layer. Your poster must have the following components: (1) a written slogan, (2) a drawing or other picture, (3) lots of color. The poster must *warn* or *inform* about the ozone layer and the potential dangers if it gets too thin.

1. Form groups as instructed by your teacher.
2. Select one person to get the art supplies, one person to use his or her book or class notes, and one person to act as a timer.
3. Get the supplies, the book, and notes and bring them back to your seats.
4. You will have thirty-five minutes to complete this task.

When the time period is up, your group will *all* hold up your poster in the front of the classroom.

Scribe's Name ______________________________ Date ____________________

18. Problems in Space Travel

In *Star Trek* and other space movies, space travelers seem to have no more problems in their daily lives than you do in your home. While these might be more problems than you want to deal with on a regular basis, the implication is that being out in space is "just like home." However, this is not the case when our real space voyagers go beyond the earth's atmosphere.

1. List all of the problems you can think of if you were going to travel in space. This list should be as complete as you can make it, based on what you now know about the conditions in space and our current technology.

2. Rearrange your list into categories. Examples of categories to use might be "Physiological Problems," "Fuel Problems," and "Time Problems."

3. Throughout this unit, add to your lists as you think of new problems. You may continue to add problems until the date this assignment is due in class.

Note: You should have a minimum of ten problems. If you did not think of ten problems, you need to think some more.

Scribe's Name ______________________________ Date ____________________

18. Problems in Space Travel *(continued)*

In-Class Assignment—Day 1

1. Get into groups as instructed by your teacher.
2. Combine all of the student lists in your group into large lists of problems under each category.
3. Based on a group discussion, come to consensus on possible solutions for each of the problems on your group lists.
4. Each group member should now select one problem and write out a detailed solution for this problem. At the top of your paper, write out the problem you have chosen. You will need to use more than one reference book or magazine to complete this step. List the author, title, and date of publication for each reference at the end of your solution.

Scribe's Name ______________________________ Date ____________________

18. Problems in Space Travel *(continued)*

In-Class Assignment—Day 2

1. Get into groups as instructed by your teacher. These groups may be the same groups as before or may be based on the problem you chose to find a solution for. Listen to your teacher.

2. In alphabetical order, each group member will explain his or her solution to the members of your group.

   a. As a solution is read aloud, each of the listening group members should write down other possible parts to the solution in addition to those given by the reader.

   b. After all solutions have been read, each person should modify his solution to include the additional suggestions given by the listening group members.

3. At the end of the reading and modifying, the group will select one of the solutions to present to the class together.

Scribe's Name ______________________________ Date ____________________

19. Death of the Sun

All living things on earth depend on sunlight for their existence. The consistent rising and setting of the sun each day and its steady glow and warmth have provided a dependable environment for life on earth to develop. In fact, life on earth has not changed much in the last three billion years, since it first developed. But what would happen to earth and its life forms if the sun changed? The sun is, after all, a star, and stars have life cycles too. What changes might occur on earth when the sun begins to die?

Below is a list of words to use in the writing of a story detailing changes in the dying sun and its effects on the earth and human life. The story should have one of two endings: (1) mankind is destroyed, (2) mankind survives. You must use all of the words on the list to receive complete credit. You can use any of the words more than once, and there are many words not on the word list that could be appropriate for this story.

ATMOSPHERE
CROPS
DESERT
EARTHQUAKES
EXPAND
FARMLAND
FLOODING
GALAXY
GIANT
INTERSTELLAR TRAVEL
LIGHT
MELT
MERCURY
OCEANS
PERISH
POLAR ICE
RED GIANT
STEAM
TEMPERATURE
VACUUM
VAPORIZE
VENUS
VOLCANOES

(continued)

Scribe's Name ______________________________ Date ____________________

19. Death of the Sun *(continued)*

You are to pretend that the sun is actually dying. Write your idea of what will happen. When you use one of the words from this list in the story it must be in ALL CAPITAL LETTERS and <u>underlined</u>. Example: <u>RED GIANT</u>. For you to receive full credit, some of the <u>CAPITALIZED</u>, <u>UNDERLINED</u> words must appear in the correct order that they would occur as the sun dies.

Write a rough draft in pencil and staple it to the back of your final draft. Your final draft should be written in ink or typewritten using as many pages as necessary. Treat this assignment like an English writing assignment. Proofread your final draft to be sure that you have no errors in spelling or grammar.

Scribe's Name ________________________________ Date ______________________

19. Death of the Sun *(continued)*

In-Class Student Follow-up

On a separate sheet of paper, answer the following questions about the process of the death of the sun.

1. How many years will it be until the sun starts to die?
2. How long has our sun been shining?
3. How will the sun change in size and color when it first starts to die?
4. Which two planets will be the first to disappear as the sun dies
5. What will happen to the earth's atmosphere as the sun dies?
6. Planet-wide flooding on earth would be caused by what occurrence associated with the sun's death?
7. When a star expands and changes color, what name is it known by?

Scribe's Name ______________________________ Date ____________________

20. An Interview with Galileo

For centuries people believed that the earth was the center of the universe, that the sun and all of the other planets traveled in circular orbits around the earth. This view was most completely described by a Greek astronomer named Ptolemy, who was alive in A.D. 120. Ptolemy's explanation was so good (because it helped astronomers predict where stars, etc., would be in the sky at different times of the year) that the idea of an earth-centered universe was accepted and believed for nearly 1500 years!

In the early 1500's a Polish mathematician had been studying the stars and didn't think Ptolemy's explanation was all that good. He thought the positions of the heavenly bodies could be much better explained if we lived in a **solar system**—if the sun was the center of the universe, and the earth and the other planets traveled in perfect circular orbits around it. Nicolaus Copernicus knew this idea wasn't going to be very popular with some religious people because the Catholic Church had always believed that the universe centered around the earth. Copernicus wasn't very interested in publishing his new theory and starting a lot of arguments, so he waited awhile to publish it. In fact, he waited until just before he died, in the year 1543.

About twenty years after Copernicus died, a person named Galileo Galilei was born. Galileo was an Italian, a genius, and an inventor. He had a tremendous impact on science, and he got into really big trouble with the Catholic Church. Do some research (a minimum of two sources please) and tell the story of Galileo's life in the form of an interview. You are the interviewer and may ask Galileo questions, which he will reply to. Include all the juicy tidbits you can find, but also include the answers to all the questions listed below.

Questions

1. What was Galileo interested in as a young boy—what kinds of things did he like to do?
2. In his twenties, Galileo observed something hanging from a cathedral tower. What was it, and what important ideas did he come up with after seeing it?
3. What are two things Galileo invented?
4. Why was he dropping weights from the Leaning Tower of Pisa, and what did this experiment do for his teaching career?
5. What discoveries did he make about the moon?

(continued)

Scribe's Name ______________________________ Date ____________________

20. An Interview with Galileo *(continued)*

6. Galileo said that something was "so numerous as to be almost beyond belief." What was he talking about?
7. What discovery did he make that had something to do with four moons?
8. Much of the work that Galileo did upheld the theories of what earlier scientist?
9. Why was the Catholic Church so irritated with Galileo, and what did the Church eventually do to him?
10. What was *Dialogues on the Two New Sciences*?
11. In 1687 another famous scientist used Galileo's *Dialogues* as a basis for his work. Who was this famous scientist?
12. How did Galileo spend the last few years of his life?

Here is an example of an interview, but don't you be this boring!

Interviewer: Good morning Mr. Galilei. Thank you so much for agreeing to this interview, it's such a thrill for me! Tell me please, in what year were you born?

Galileo: 1564.

Interviewer: And what year did you die?

Galileo: 1642.

Interviewer: And what did you invent?

(etc.)

Write a rough draft in pencil and staple it to the back of your final draft. Your final draft should be written in ink or typewritten using as many pages as necessary. Treat this assignment like an English writing assignment. Proofread your final draft to be sure that you have no errors in spelling or grammar.

Scribe's Name ______________________________ Date ____________________

20. An Interview with Galileo *(continued)*

In-Class Student Follow-up

1. Get into groups as instructed by your teacher.

2. Get out one piece of smooth-edged notebook paper and put all group members' names on the top.

3. Develop a group consensus answer for each of the following questions. Write the consensus answers on the smooth-edged notebook paper.

   a. What was Galileo's greatest contribution? Why?

   b. What did the Tower of Pisa experiment demonstrate?

   c. How did Galileo's explanation of moonlight differ from Aristotle's?

   d. How did Galileo's research influence Isaac Newton?

   e. Give an example of an issue in today's society that is as controversial as Galileo's sun-centered planetary system. Explain both sides of this controversial issue.

4. Listen carefully while each person reads his or her dialogue aloud.

5. After each person reads his or her dialogue aloud, make sure that the dialogue provides a complete answer to each question in the assignment. If the group decides that something needs to be added to the dialogue, write the new part on the back of one of the pages.

6. Put all dialogue parts together in the order your group agreed to use.

7. If any of the parts seem out of place, change the order to form the best possible group dialogue. The group may add small snatches of dialogue between the individual parts if it makes the group dialogue sound more natural.

8. Make a stack of dialogues in the best order.

9. Staple the group consensus answers to the top of your group's stack of dialogues. Turn the completed stack in to your teacher.

(continued)

Scribe's Name ______________________________ Date ____________________

21. Story of a Star

For centuries people have watched the night sky and wondered at the vastness of the universe and the steady glow of the stars, and taken comfort in the unchanging view above. Confirmed stargazers may track the movement of the stars throughout the year, but most people see no changes in the stars from one evening to the next (or in our closest star, the sun, from day to day). It is hard to believe that the stars above are in a constant state of intense chemical activity and that some stars are being born while others are dying, with lifetimes that last for billions of years.

Despite differences in chemical makeup, size, and location in the universe, all stars go through much the same kind of life span. Write a story in which you describe the birth, life, and death of a typical star. Below is a list of words to use as a guide in writing this story. You must use all of the words in your story for complete credit. You can use any of the words more than once, and there are many words not on the word list that could be necessary or appropriate for this story.

ATOMS
BLACK HOLE
COLLAPSE
DENSE/DENSITY
DUST
ENERGY
EXPAND
EXPLOSION
FUEL
FUSE/FUSION
GAS
GRAVITY
HELIUM
HOT/HOTTER
HYDROGEN
LIGHT-YEARS
MATTER
MILLIONS
NEBULA/NEBULAE
NEUTRONS
NUCLEAR
PULSAR
RADIO WAVES
RED GIANT
SECOND
SHRINK/EXPAND
SUPERNOVA
TEMPERATURE
WHITE DWARF
YELLOW

(continued)

Scribe's Name ______________________________ Date ____________________

21. Story of a Star *(continued)*

Pretend that you are a star. Write your autobiography. You may assume that you are a very mature (old) star, but you have a great memory of past events. Be sure to include each of the stages in a star's life in your story. When you use one of the words from this list in the story, it must be in ALL CAPITAL LETTERS and underlined. Example: SUPERNOVA. Some of the CAPITALIZED, UNDERLINED words must appear in the correct order that they would occur in the process to receive full credit.

Write a rough draft in pencil and staple it to the back of your final draft. Your final draft should be written in ink or typewritten using as many pages as necessary. Treat this assignment like an English writing assignment. Proofread your final draft to be sure that you have no errors in spelling or grammar.

Scribe's Name ______________________________ Date ____________________

21. Story of a Star *(continued)*

In-Class Student Follow-up

Answer the following questions about the process of the death of the sun.

1. What is a **nebula**?
2. What is the name for an incredibly dense, hot star?
3. A star's lifetime may last how many years?
4. A star's light and heat come from what process?
5. What are the reactants and what is the product in a star's **fusion** process?
6. As a star shrinks, what happens to its temperature?
7. Put these stages in a star's life cycle in order from first-occurring to last: **black hole**, **nebulae**, **pulsar**, **red dwarf**, **star**, **supernova**, **white dwarf**.

Scribe's Name ____________________ Date ____________

22. Lost in Space

An Activity in Recognizing and Analyzing Relevant Data

You are a crew member aboard the U.S. Spaceship *Discovery* on its mission to explore the outer reaches of our solar system. *Discovery* is prepared for a mission of incredible length, perhaps hundreds of years. Consequently, all crew members are cryogenically "frozen" and have nothing to do with the day-to-day running of the spaceship; this is all under computer control. Unfortunately, *Discovery*'s computer has malfunctioned and the ship has veered wildly off course. Since all crew members are unaware of the malfunction, no correction has been made and *Discovery* has traveled in space for an unknown amount of time in an unknown direction.

Until now, that is—*Discovery* has just crash-landed on a planet within our solar system. Damage to the spaceship is major. All instrumentation is broken, set at the reading taken just minutes before the crash occurred. Worse still is the condition of the crew. All the cryogenic pods, save one, have been ruined, and the crew members contained within are dead. You are the sole survivor, automatically "thawed out" and returned to normal functioning.

As you look out one of *Discovery*'s portholes you see the sun. The sun appears to be larger in the sky than you remember it to be from Earth. The apparent movement of the sun across the sky is very slow; in fact, the sun appears to have moved only slightly since yesterday's crash. Compass readings indicate the sun will eventually set in the east. Looking out another porthole, you see pale yellow clouds in an otherwise clear white sky. There is a very high, very heavy cloud cover that obstructs the view of your planetary neighbors.

The landscape you see is rocky, rolling plains that glow with a reddish color. There are no apparent signs of life.

Instrument readings show that the outside temperature is over 800 degrees Fahrenheit. A chemical analysis of the planet's atmosphere available from the ship's instrumentation indicates a high percentage of carbon dioxide gas and sulfuric acid. The atmospheric pressure on this planet is approximately 2,200 pounds per square inch.

If you are to be rescued, you must determine where you are. Information available to you includes the instrument readings taken just before the crash, what you can see through *Discovery*'s portholes, and a set of data charts on the solar system's nine planets from the ship's library (also to be found in your local or school library). Please complete the log book page you were given with the correct information.

(continued)

Scribe's Name ______________________________ Date ____________________

22. Lost in Space *(continued)*

UNITED STATES SPACESHIP DISCOVERY *CREW MEMBER'S PERSONAL LOG*

1.0 Earth Date: October 15, 2116

2.0 Name of Crew Member: ______________________________

3.0 Personnel Identification Number: ______________________________

4.0 Rank and Duties: ______________________________

5.0 Ship's Mission and Status at Time of Log Entry: ____________________

6.0 Ship's Location:

6.1 Galaxy: ______________________________

6.2 Star System: ______________________________

6.3 Planet: ______________________________

7.0 Observations:

7.1 Observations That Support Planet Location:

A. ______________________________

B. ______________________________

C. ______________________________

D. ______________________________

E. ______________________________

F. ______________________________

7.2 Explanation of Observation:

A. ______________________________

B. ______________________________

C. ______________________________

(continued)

Scribe's Name ______________________ Date ______________

22. Lost in Space *(continued)*

D. ______________________________

E. ______________________________

F. ______________________________

8.0 Approximate Distance from Center of Star System: ______________

9.0 Approximate Diameter of Planet: ______________

10.0 Nearest Planetary Neighbor in Star System: ______________

11.0 Verification of Location Based On: ______________

____

____

____

____

(Use a minimum of two references: List the author, title, and date of publication for each reference.)

12.0 Other Observations:

____

____

____

____

____

____

Scribe's Name ______________________ Date ______________

23. How Many *Elis* Tall Are You?

Personal Equivalency Assignment

Let's say you had enough money to buy one large pizza. However, some of the friends you are with would rather have burgers than pizza. Quickly you calculate how many burgers you can afford. You announce that you could buy seven burgers. In this situation, one pizza is equivalent to seven burgers.

In this exercise you will be asked to determine your height, head size, foot size, and seat distance from one point in the center-front of the room.

1. Make a chart with the headings below:

Name	Height (cm)	Head Size (cm)	Foot Size (cm)	Distance from Front (cm)

   Record each of your classmates' data in these columns.

2. You will now determine these four sizes and distances again. However, you will not be using inches, centimeters, feet, or loafer measurements. You will figure out these four personal characteristics in "classmate equivalencies." In other words, you will figure out how tall you are in Classmate #1's, how big your head is in Classmate #2's, how big your foot is in Classmate #3's, and how far you are from the front in Classmate #4's. If the name of classmate #1 was Eli, your height would be in *Elis.* If classmate #2 was Kristina, your head size would be in *Kristinas,* etc.

3. Listen carefully to your teacher to find out which of your classmates will be the equivalency standard for each of the characteristics.

4. To figure out how many *Elis* tall you are, divide your height by Eli's height. Round this number off to one decimal point.

5. Record your personal equivalencies in a chart like the one you have already made. This second chart will only have your *own* equivalents (*Elis,* etc.).

Name	Height	Head Size	Foot Size	Distance from Front
Me	1.25 *Elis*	.95 *Kristinas*	Etc.	Etc.

(continued)

Scribe's Name ______________________ Date ______________

23. How Many *Elis* Tall Are You? *(continued)*

Earth Equivalency Assignment

Let's say you had enough money to buy one large pizza. However, some of the friends you are with would rather have burgers than pizza. Quickly you calculate how many burgers you can afford. You announce that you could buy seven burgers. In this situation, one pizza is equivalent to seven burgers.

Procedure

1. The chart below is of the average distances, times, or sizes of planet's orbits or diameters.
2. You will be given a scale to use to calculate the number of Earth equivalents involved in one of the sizes or times for each planet.
3. Your job is to make a chart like the one below, and calculate the number of Earth equivalents involved for each planet for each characteristic. The scale will have the Earth value equal to 1 cm, 10 cm, or 1 m. Check with your teacher to see which scale you will use.
4. After you have determined the number of Earth equivalents, determine the exact size, time, or distance each planet should be assigned based on your scale.

Planet	Dist. from Sun[1]	Diameter[2]	Revolution[3]	Rotation[4]
Mercury	58	4,878	88	1,408
Venus	108	12,104	226	5,832
Earth	150	12,756	365	24
Mars	228	6,787	686	25
Jupiter	778	142,800	4,329	10
Saturn	1,427	120,000	10,753	11
Uranus	2,870	51,200	30,664	17
Neptune	4,497	49,427	60,148	16
Pluto	5,900	2,300	90,407	153

Notes:
[1]Distance given in millions of kilometers
[2]Diameters given in kilometers
[3]Revolutions given in Earth days
[4]Rotations given in Earth hours

(continued)

Scribe's Name ______________________ Date ______________

23. How Many *Elis* Tall Are You? *(continued)*

Example:

If your Earth equivalent for the diameter of Pluto was

$$\frac{2300}{12{,}756} = 0.18$$

and the scale you were using was

"Earth = 1 inch,"

then Pluto would be

.18 (times) 1 inch or .18 inches in diameter (about $\frac{1}{8}$ inch).

The general formula is (Earth equivalents) × (scale) = (size or time).

Scribe's Name ______________________________ Date ____________________

24. A Trip to My Favorite Planet

When you get ready to take a vacation, you might visit a travel agent to help plan air or ground travel, to reserve a hotel room, or even to pick the ideal vacation spot. Because of this, the walls of travel agencies are covered with beautiful posters of exotic vacation locations.

Your task in this exercise is to design a colorful travel poster for one of the planets in our solar system. The poster should be no larger than 11 × 17 inches. It should have at least one large picture of the planet or one of its features. There should be between three and five slogans on the poster. Each slogan should describe in words some aspect of the planet or some make-believe amusement-parklike ride that might be found there. Each slogan should be inside some design, like the words of a cartoon character or the sound when something explodes in a comic strip. The name of the planet should be at the very top or bottom, or down the side of the poster in large letters.

Check with your teacher to see if you can choose any planet you want or if you have to sign up for a specific planet. Check also to see if there is a short written portion to this assignment.

Scribe's Name ______________________ Date ______________

25. "The Creature That Came from . . ."

For as long as there have been science fiction writers, there has been the concept of creatures that live on real or imaginary planets. Consider this description:

> *Here were the great males towering in all the majesty of their imposing height; here were the gleaming white tusks protruding from their massive lower jaws to a point near the centre [sic] of their foreheads, the laterally placed, protruding eyes with which they could look forward or backward, or to either side without turning their heads, here the strange antennaelike ears rising from the tops of their foreheads; and the additional pair of arms extending from midway between the shoulders and the hips.*
>
> —Edgar Rice Burroughs, *The Gods of Mars*. Chapter I: "The Plant Men" (1912)

In his series of ten Martian novels, Burroughs describes five complete civilizations and a host of strange creatures that occupy the slowly dying red planet. The creatures described above belong to a race of green men who make up one fourth of the planet's main inhabitants.

Burroughs's Martians all had significant modifications that allowed them to survive in the lower gravity and thin atmosphere of Mars. Of course, recent scientific discoveries have shown that there is no oxygen on Mars, nor enough water to allow life as we know it to exist. Sometimes science takes some of the fun out of imagination, *but not today*!

Today your assignment is to invent a creature for one of the planets described below. Your creature should be adapted to the conditions given for the planet you select or are assigned. After you invent your creature, name it. Draw a picture of your creature. Select three features of your creature and explain why you gave the creature each feature. Write one to two paragraphs of explanation that refer to two of the conditions from the list for the planet. Do the write-up as if you are creating a commercial trying to get humans living on the planet to buy your creature as a pet, or warning visitors to stay away from your creature when they visit the planet.

1. Form groups as instructed by your teacher.

2. Each member of the group should select one of the planets from the following table to be the planet where their creature lives. Only one group member can do any one of the planets.

(continued)

Scribe's Name ______________________ Date ______________

25. "The Creature That Came from . . ." *(continued)*

3. Write down each group member's name and which planet they chose and turn that paper in to your teacher.

	Mercury	**Venus**	**Mars**	**Jupiter**	**Uranus**
Day Length[1]	1,408	5832	25	10	17
Temperature Extremes[2]	475°C/–170°C	475°C/–33°C	-31°C/–130°C	29,700°C/–95°C	?/–370°C
Atmosphere[3]	None	CO_2	CO_2	H_2, He	H_2, He, Methane

[1]Day length is given in Earth hours.

[2]Temperature extremes are given as best estimates of high temperature and low temperature on the planet.

[3]Atmosphere is given as major components derived from fly-by and spectroscopic data.

Use your creative ability. Your picture should be your very best art work. Colors and any background scenery will only enhance the feeling of reality.

Write a rough draft in pencil and staple it to the back of your final draft. Your final draft should be written in ink or typewritten using as many pages as necessary. Treat this assignment like an English writing assignment. Proofread your final draft for errors in spelling or grammar.

Scribe's Name ______________________________ Date ____________________

25. "The Creature That Came from . . ." *(continued)*

In-Class Student Follow-up

1. Form groups as instructed by your teacher.
2. Listen while each group member reads his or her written assignment.
3. Look at the picture each group member drew.
4. Compare each group member's written description with his or her picture.
   a. If all the written descriptions match the picture, give the paper one star.
   b. If each written description explains one of the three characteristics of the planet, give the paper one star. If each written description explains two of the three characteristics of the planet, give the paper two stars.
   c. If the picture is neat and in glorious color, give the paper one star.
   d. If there are no spelling or grammar errors in the written descriptions, give the paper one star.
   e. If the picture has background scenery, give the paper one star.
5. Write down group consensus answers to the following questions:
   a. What is the best way for a creature to adapt to extreme heat?
   b. What is the best way for a creature to adapt to extreme cold?
   c. What is one difference between your respiratory system and the respiratory system of a creature on Uranus?
   d. The gravity of Mars is about one sixth Earth's gravity. What would a Martian have to have to be able to adapt to the lesser gravity? Give an example of one thing a Martian could do on Mars that you could not do here on earth.
   e. What would you eat if you were a creature on Jupiter?
   f. What adaptations would you have to have if you were a plant on Venus?
6. Staple all of the group papers together and turn them in to your teacher.

Scribe's Name ______________________________ Date ____________________

26. That Really Burns Me Up

Jenny and Lynda were spending a day at the beach. Most of their time was spent people-watching and soaking up the sun. They still had not been in the water, and it was after 1:00. But there was no time for swimming now. It was serious tanning time. Both girls got out their bottles of tanning product and wiped down their arms, legs, and torsos.

They burst into giggles after turning to ask each other to put lotion on their back. Then, without thinking about what they were doing, they rubbed their own lotion onto the other's back. After a lingering stare at a great-looking guy walking down the beach, they stretched out on their stomachs and dozed off.

It was over an hour later when Lynda awoke with a start. She checked her watch and shook her friend to awaken her.

"Jenny, it's nearly 3:00!" she cried. "My Mom's gonna be here any second!"

In record time the two girls had packed their stuff and were standing on the boardwalk waiting for their ride home.

At 10:00 the next morning Jenny's phone rang.

"Jenny, what kind of lotion did you have in your bag?" Lynda moaned. "My back is redder than a cooked lobster!"

"Your back hurts?" Jenny asked. "That's the only place I don't hurt. What was in your bottle?"

"My lotion says 'SPF 35'."

"What does that mean?"

"Uhh, let me look. Here it is. It says, 'This lotion provides a sun protection factor of 35. Waterproof for 90 minutes.' I think it means that you can stay in the sun 35 times longer than you normally would without burning."

"Hold on," Jenny said. "Let me get my lotion. Mine doesn't say anything about sun protection. That must be the difference. I wonder how that stuff works. Ooohh, I wish I had used your lotion all over. I've never been so burned!"

* * * * *

(continued)

Scribe's Name ______________________________ Date ____________________

26. That Really Burns Me Up *(continued)*

Situations like the previous one happen all the time. It might have happened to you. In fact, there is a very good chance that you have been sunburned at some time or another. Until the 1980's, the extent of the negative effects of suntanning and burning were not known.

Today, the type of permanent damage that is done to your skin by **ultraviolet light** is known. People are strongly encouraged to cover up if they must be outside. If covering up is not possible, people are cautioned to protect their skin with lotion containing an ultraviolet sun protection ingredient. Sometimes called **sunblock**, or **sunscreen** this essential ingredient actually blocks the ultraviolet rays of the sun and prohibits them from penetrating and damaging the DNA of the cells beneath the surface of your skin for a time that varies from person to person and with the amount of lotion used.

Your task in this assignment is to write one fourth of a group dialogue between a **melanocyte**, one of the pigment-producing cells in your skin, and your brain. This should look and sound like a play when you have finished. (See example below.) The melanocyte's job in this dialogue is to convince the brain of the damaging effects of sunlight on the skin. You will need to use reference books or textbooks to get specific information to help you address one of the following problems: (1) what ultraviolet radiation is, (2) the structure of the skin, (3) the structure and function of DNA, or (4) the damage, both temporary and permanent, that ultraviolet radiation does to the skin.

Melanocyte: Hey! Give me a break, will ya?

Brain: What? Say, who's calling to me now? I mean, I've got all kinds of important functions to do, not counting daydreaming about—

Melanocyte: I know you've got it tough, Brainy. But I'm dying out here!

Brain: Is that you, Mel? How are things out on Sunbaked Shoulder?

Mel: Sunburned Shoulder, you mean. Let me tell you . . .

(continued)

Scribe's Name ______________________ Date ______________

26. That Really Burns Me Up *(continued)*

Procedure

1. Get into groups as instructed by your teacher.
2. Decide which person in your group is going to research and write about which of the four problems (ultraviolet radiation, structure of the skin, structure and function of DNA, or temporary and permanent damage to the skin from ultraviolet radiation).
3. Decide in what order your group dialogue should include each of the problems.
4. Write down the problems in the order they will be included in the team dialogue. Put the name of the group member who is researching each problem beside the problem. Turn this list in to your teacher.

Use your creative ability. Write a rough draft in pencil and staple it to the back of your final draft. Your final draft should be written in ink or typewritten using as many pages as necessary. Treat this assignment like an English writing assignment. Proofread your final draft for errors in spelling or grammar.

Bring your section of the dialogue with you to class on the day this assignment is due.

Scribe's Name ______________________________ Date ____________________

26. That Really Burns Me Up *(continued)*

In-Class Student Follow-up

1. Get into groups as instructed by your teacher.

2. Get out one piece of smooth-edged notebook paper and put all group members' names on the top.

3. Develop a group consensus answer to each of the following questions assigned by your teacher. Write the consensus answer on the smooth-edged notebook paper.

   a. What do adenine, **guanine**, **thymine**, and **cytosine** have to do with this assignment?
   b. Where in the skin is the dermis? Name four structures found there.
   c. What is the **electromagnetic spectrum**?
   d. What is the first sign of ultraviolet damage to the skin?
   e. What part of the skin tries to protect skin from ultraviolet damage? How does it try?
   f. Why is ultraviolet light such a problem to the skin?
   g. What does normal, undamaged DNA do in your cells?
   h. Explain how skin cancer can result from long-term exposure to sunlight.
   i. How does decreasing the ozone layer affect this situation?
   j. What is the best SPF number to have in a sunscreen product? Why?

4. Listen carefully while each person reads his or her dialogue aloud.

5. After each person reads his or her dialogue aloud, make sure that the dialogue provides a complete answer to the problem that person was assigned. If the group decides that something needs to be added to the dialogue, write the new part on the back of one of the pages.

6. Put all four dialogue parts together in the order your group agreed to use.

7. If any of the parts seem out of place, change the order to form the best possible group dialogue. The group may add small snatches of dialogue between the individual parts if it makes the group dialogue sound more natural.

8. Make a stack of dialogues in the best order.

9. Staple the group consensus answers to the top of your group's stack of dialogues. Turn the completed stack in to your teacher.

Scribe's Name ______________________________ Date ____________________

27. One Step at a Time

A Demonstration of Enzyme Function and Enzymatic Pathways

Enzymes are specialized proteins that function very specifically in living systems. Each enzyme is so specialized that it can only do one thing or one job. In other words, if it breaks one molecule into two molecules, it *cannot* put those two molecules back together. Because of this fact, enzymes often work in groups called **pathways**. An enzymatic pathway can do an entire task because the enzyme that does a particular step in the task is next to the enzyme that does the next step, and so on, like paving stones in a pathway.

Regardless of the pathway, one step determines how long it will take for the entire reaction to be completed. This step, the **limiting factor**, might have to do with the length of time it takes for the step to occur. Other factors that might make a step limiting are the amount of material or the amount of enzyme.

In the simulation that follows, you will assume the role of an enzyme. Your group will represent an enzymatic pathway. Your task is to complete a poster like the sample your teacher will show you.

1. Get into groups of six people as instructed by your teacher.
2. Study the poster. Decide which group member will cut out the cone, which will cut out the ice cream, which will cut out the cherry, which will cut out the sign, and which will cut out the price. Whoever is left over will be the gluing enzyme. Answer question 1.
3. Study the poster. Answer question 2.
4. Pick up the six pieces of the poster, five pairs of scissors, and glue from your teacher.
5. Write down the time you start this step as the answer to 3 below. Give each of the cutting enzymes his or her scissors and have them neatly cut out their parts.
6. As each part is cut out, the cutting enzyme should hand it to the gluing enzyme. Answer question 4.
7. The gluing enzyme should glue the parts onto the background in the order listed in question 2. Answer question 5.
8. Answer questions 6–9.

(continued)

Scribe's Name ______________________________ Date ____________________

27. One Step at a Time *(continued)*

Questions

1. Who is cutting out each part? Write down the names of each part and who is cutting it out. Then write down the name of the gluing enzyme.
2. In what order should the pieces be glued onto the background? Write down what your group thinks the correct order should be.
3. Write down the time as instructed in step 5 above.
4. Write down the time at which the last cutting enzyme handed its product to the gluing enzyme.
5. Write down the time the gluing enzyme finished gluing the last piece on the poster.
6. Which step in the process was the limiting factor? Explain why.
7. Was your order of gluing from step 2 the best one for this pathway? If no, list a better order and explain what was wrong with your hypothesized order.
8. What would have happened if the gluing enzyme had glued the parts of the poster on the background in the order he or she received them from the group? You need to list specific things that would have gone wrong.
9. Suggest a way to speed up the limiting factor step. Once again, this should be very specific.

Scribe's Name ______________________ Date ______________

28. Liver Without Onions in a Peroxide Sauce—Yum!

Liver and onions! Some people like it; most people do not. You will *not* be using onions in this exercise. You *will* be using liver and peroxide to create some nontypical situations.

1. Get into groups as instructed by your teacher.
2. Read through the entire procedure.
3. On a clean piece of smooth-edged notebook paper, do the following.
   a. Write the purpose of this lab, which you will have figured out while reading through the procedures.
   b. Reproduce the following chart on your paper beneath the purpose.

Trial No.	**Starting Temp**	**Temperature After Adding Liver**							
		.5	1	1.5	2	2.5	3	3.5	4 (minutes)
1									
2									
3									
4									
Total									
Avg.									

   c. Write down any questions you have about the procedure.

Procedures

1. Get a test tube, four pieces of liver, and a thermometer.
2. Take your test tube to the sink area.

(continued)

Scribe's Name ______________________________ Date ____________________

28. Liver Without Onions in a Peroxide Sauce—Yum! *(continued)*

3. Measure 5 ml of hydrogen peroxide into a graduated cylinder. Pour the peroxide into your test tube.

4. Return to your seat and place your test tube into the rack there.

5. Place the thermometer in the tube and take the temperature of the peroxide. Record that datum on your chart. Remove the thermometer from the tube.

6. Place one piece of liver into the tube.

7. Put the thermometer back into the tube. You might need to push the liver into the peroxide with the thermometer. If you do need to push the liver, be careful not to poke the thermometer through the bottom of the test tube. Record the temperature every thirty seconds (one-half minute) for four minutes.

8. Empty and rinse the test tube with tap water. Throw the used liver into the appropriate receptacle, not the sink! Answer questions 1, 2, and 3 below.

9. Repeat steps 2 through 8 three additional times.

10. Complete the chart.

11. Graph your average (Avg) data. **Time** is the *x*-axis. **Degrees Celsius** is the *y*-axis. Connect your data points with a line.

12. Answer questions 4–9 below.

Questions

1. What evidences (note the plural form of the word) did you have that some type of chemical change was occurring?

2. Hydrogen peroxide has the formula H_2O_2. What do you think happened to the hydrogen peroxide **molecules** during the experiment?

3. What happened to the liver? Be specific.

(continued)

Scribe's Name ______________________________ Date ____________________

28. Liver Without Onions in a Peroxide Sauce—Yum! ***(continued)***

4. During each thirty-second time period did the temperature change by approximately the same number of degrees? Why or why not?

5. What time of what trial had the highest temperature?

6. Were the other temperatures at that time in the other trials the highest in their group?

7. What made the bubbles during each trial?

8. Why did you repeat this experiment four times?

9. Would you have produced the same curve on your graph if you had graphed each trial separately? Explain.

Scribe's Name ______________________________ Date ____________________

29. Funnels of Fun

In chemical reactions, there is one step, or one component of the reaction that is the slowest part of the reaction. That step or component is known as the **limiting factor** in the reaction. While it is easy to describe the concept of limiting factor, it is not always quite as easy to understand what a limiting factor is.

In the exercise that follows, you will be doing a procedure to illustrate what the limiting factor is and how it works in determining the rate of a reaction.

1. Get into groups as directed by your teacher.
2. Select a funnel of your own. The only restriction on funnel selection is that each person's funnel should be significantly different in size from the others in your group. Label the funnel with your name. Answer question 1 below.
3. Your group will also need a beaker with 250 ml of water in it and some form of beaker or bucket in which to collect the water.
4. As rapidly as you can, *without spilling* any of the water outside of the funnel, pour 250 ml of water through your funnel. Time this procedure. The time should start when water starts to come out of the small end of the funnel. The time should stop when the water stops *running* out of the funnel. Have someone else record the time (in seconds) it takes for the 250 ml of water to run through your funnel.
5. Record the times for each person's funnel in the group.
6. Measure the internal diameter of both ends of each funnel. Record these measurements on the same line of the chart as the time it took for that funnel. Use the abbreviations B.E. for "big end," and S.E. for "small end."
7. Hypothesize the best sequence of funnels to allow for the fastest flow of water through all of the funnels consecutively. Write down the names of the funnels in this best arrangement.
8. Be sure the questions in 2 below are answered and go on to question 3 on the next page.
9. Have each person hold his or her own funnel in the order from step 7 above.

(continued)

Scribe's Name ______________________________ Date ____________________

29. Funnels of Fun *(continued)*

10. Pour the 250 ml of water through the funnels. Be sure that all of the water goes through each of the funnels and not over the sides.

11. Repeat the previous step. But, this time, record the time it takes for the water to run through the funnels.

12. Rearrange the funnels into a new sequence. Write down this sequence by the name of the funnels as you did in step 7.

13. Repeat the experiment after rearranging the funnels. Record the time it takes the water to go through.

14. Rearrange the funnels a third time and record the new arrangement. Pour the water through and record the time. Answer questions 4–6.

Questions

1. Make a table like the one which is shown below.

Name	Time	B.E.	S.E.

2. a. Write the names of each funnel in the first column.

   b. Write down the times it took for the 250 ml of water to go through each funnel.

   c. Record the internal diameter of both ends of each funnel on the same line as the time it took for that funnel.

(continued)

Scribe's Name ______________________________ Date ____________________

29. Funnels of Fun *(continued)*

3. Graph the relationship between both the size of the small end and the large end of each funnel separately against the time it took for the water to go through. Time (in seconds) should be the *y*-axis of the graph.

4. a. List the arrangement of funnels that provided the fastest time for the water to flow through.

   b. List the slowest arrangement.

5. The time to go through all the funnels was closest to (maybe equal to) the time for which single funnel?

6. What is the limiting factor in this experiment? Explain.

Scribe's Name ______________________________ Date ____________________

30. Sometimes You Feel Like a Nut

One of the best ways to lose weight is to "burn" calories. You can burn calories by exercising and you can feel those calories burning because you get hot when you exercise. In fact, a calorie is a measure of how much heat energy is contained in a food. However, when your body burns calories, it is pretty hard to measure how many came from which food you just ate. There's an easier way to determine the heat energy contained in a food.

Read this laboratory procedure and write down the purpose on your answer sheet.

Procedure

1. Get into groups as directed by your teacher. Write your group number on your answer sheet.
2. On your answer sheet, make a copy of the chart below. Do not write on this paper.

Group Number	Type of Nut	Weight of Nut	Temp After Nut Burned	Temp Before Nut Burned	Net Change in Temp	Calories Absorbed	Calories Released	Cal/g in Nut
1 . . . 12								

3. Set up your equipment so it looks like the teacher's demonstration setup.
4. Go and get a nut from your teacher.
5. Weigh your nut in grams. Record this on your chart.
6. Carefully push your nut onto your straight pin.
7. Pour 50 ml of water into your flask.
8. Place the flask on the ring stand.

(continued)

Scribe's Name ______________________________ Date ____________________

30. Sometimes You Feel Like a Nut *(continued)*

9. Take the temperature of the water. Record this number in the "Temp Before Nut Burned" column on the chart.

10. Gently slide your coffee can under the setup.

Demonstration Setup

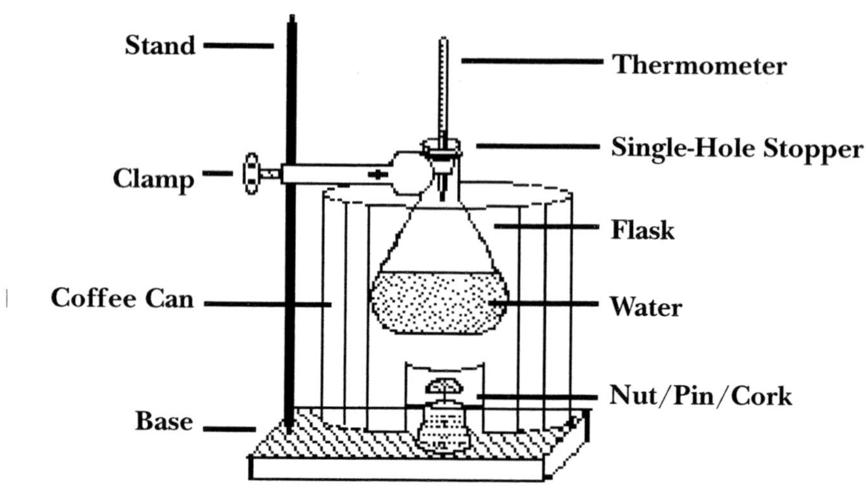

11. Use a match to set the nut on fire.

12. Slide the nut under the ring stand with the flask on it.

13. Watch the nut burn. Answer questions 1 and 2 below.

14. After the nut stops burning, take the temperature of the water. Record this number in the "Temp After Nut Burned" column on the chart.

15. Listen to your teacher to find out how to clean up and when you may exchange information with the other groups in order to complete your data table.

16. Answer questions 3 through 10 below.

Questions

1. What is in the nut that is burning?

2. What is the purpose of the coffee can?

(continued)

Scribe's Name ______________________________ Date ________________

30. Sometimes You Feel Like a Nut *(continued)*

3. For your nut, show how you calculated the change in water temperature.

4. The formula for determining the number of heat calories absorbed by the water is:

   (Mass of Water) × (Temp Change of Water) × (1 calorie per gram of water) = calories absorbed

   Fill in the above formula for your nut.

5. For your nut, calculate the number of food calories (kilocalories) released.

   The formula is $\frac{\text{calories absorbed}}{1000}$ = calories released. Fill in the preceding formula for your nut and record the answer in your data table.

6. For your nut, calculate the food calories released per gram of nut. This answer will be in calories per gram of nut (Cal/g). Record this number in your data table.

7. Which group's flask of water *absorbed* the most calories? What kind of nut did they burn?

8. Which group's nut *released* the most calories per gram of nut? What kind of nut was it?

9. Which group's nut released the *fewest* calories per gram of nut? What kind of nut was it?

10. Compare the calories in a dry roasted peanut and a regular peanut. Which type of peanut has the most Cal/g? Why do you think this is so?

Scribe's Name ______________________________ Date ____________________

31. Faster Than a Speeding . . . Snail?

On a piece of smooth-edged notebook paper write down the following numbers for answers: 1A, 1C, 2C, 3A, 3C, 4C, 4F, 4G, 4H, 5B, 6A, 6B, 6C, 6D, 7A, 7B. Be sure to skip one line between each number.

Procedure

1. First determine the weight and length of your snail.

   a. Weigh your snail on the scale provided.

   b. Record the weight of your snail on your answer sheet in space 1A.

   c. Measure the length of your snail's shell with a 15-cm ruler.

   d. Record the length in space 1C.

2. Now determine how a snail moves.

   a. Put your snail on a glass plate.

   b. Watch from the bottom of the plate while the snail crawls.

   c. In answer space 2C, describe how the snail moves.

3. Next, determine the sex of your snail.

   a. The snail is a **hermaphrodite**. What does that mean?

   b. Record your answer in space 3A on the answer sheet.

   c. In answer space 3C, record how a snail is different sexually from a human.

4. You will now determine how fast your snail can crawl.

   a. Draw a line 15 centimeters long on the back of your answer sheet.

   b. Place the tip of your snail at one end of the line.

   c. Time how long it takes the snail to crawl to the other end of the line. Use the second hand on the clock.

   d. Record this time, *in seconds*, on your answer sheet in space 4C.

   e. Do these steps 4A–4D two more times. Record these numbers in space 4C, too. Put a comma after each number.

(continued)

Scribe's Name ______________________________ Date __________________

31. Faster Than a Speeding . . . Snail? *(continued)*

f. Figure the average for the three times you raced your snail. Record the average as answer 4F.

g. Now, use the math formula below to figure out how fast your snail can move in meters per hour (mph). Answer in space 4G. Show your work.

Math formula:
(Put the answer you got in space 4C where the ? is and do the math.)

$$\frac{15\,cm}{?\text{sec}} \times \frac{3600\text{sec}}{1\,hr} \times \frac{1\,m}{100\,cm} = \text{your answer in mph}$$

h. An average snail can move 3 mph. How does your snail compare to an average snail's pace? Is it faster, slower, or the same? Answer in space 4H.

5. Now it is time to see how your snail eats. After all, it has worked pretty hard.

   a. Feed your snail a piece of lettuce.
   b. Watch it carefully and record what happens in answer space 5B.

6. Take a look at the snail's antenna.

   a. How many antenna does your snail have? Answer in space 6A.
   b. Which antennae are longer? Why? Answer in space 6B.
   c. Where are the eyes located? How does the snail protect its eyes? Answer in space 6C.
   d. Hold your finger directly in front of your snail. Watch the antennae and record what happens in answer space 6D.

7. Look at the opening of the snail's shell. You should be able to see a small breathing opening. Count the number of times it opens in one minute.

   a. Record the number of breaths per minute in answer space 7A.
   b. Humans breathe about 17 times per minute when resting. Who breathes faster, a resting human or a snail? Answer in 7B.

Scribe's Name ______________________________ Date ____________________

32. Eddie (or Edie) Electron

An electric current is often described like water: Water flows downhill; an electric current flows from a point where there are a lot of **electrons** to a point where there are fewer electrons. The path followed by the electrons is called the **circuit**. An electrical circuit can consist of many different devices—some of these are listed below. Using a physical science textbook or some other appropriate reference, define each of these parts of an electrical circuit. Tell what each part is used for in a circuit and what happens to electrons as they pass through it.

battery	**negative terminal**	**positive terminal**
insulation	**wire**	**resistance**
switch	**light bulb** (a source of resistance)	

Below is a diagram of a circuit containing all of the different devices listed above.

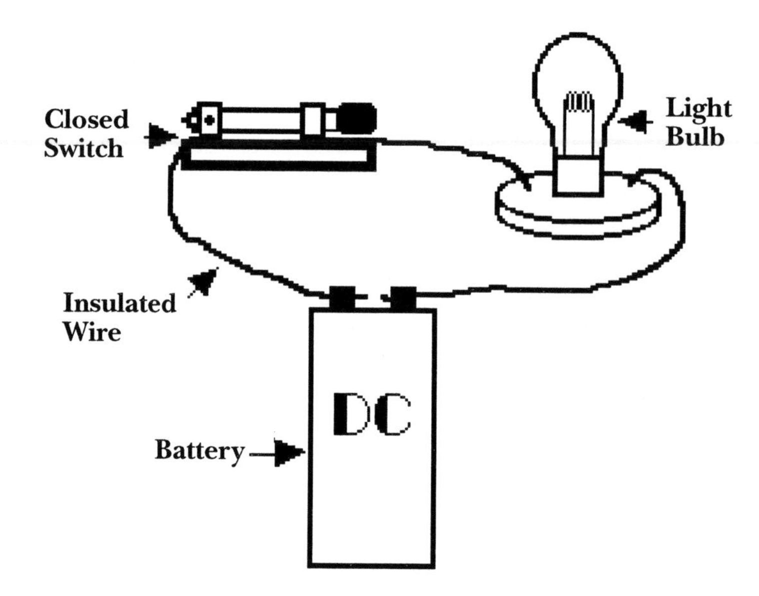

(continued)

Scribe's Name ______________________________ Date ____________________

32. Eddie (or Edie) Electron *(continued)*

Your next job is to write a story from the viewpoint of "Eddie" or "Edie" Electron. What happens to Eddie/Edie (you) as he or she moves through each of the parts of the circuit in the order shown in the diagram? Assume you are starting at the negative terminal of the battery. In which direction would you, as an electron go, and why? What happens to Eddie/Edie and his or her electron friends as they pass through each device? Be sure to include in your story all of the information you used to answer the questions above. When you use one of the words you defined in the story it must be in ALL CAPITAL LETTERS and underlined. Example: BATTERY. You may need to use some words more than once to completely describe the electron's path, but, capitalize and underline each word *only* the first time that you use that word. Some of the CAPITALIZED, UNDERLINED words must appear in the correct order that they would occur in the pathway of the electrons to receive full credit.

Write a rough draft in pencil and staple it to the back of your final draft. Your final draft should be written in ink or typewritten using as many pages as necessary. Treat this assignment like an English writing assignment. Proofread your final draft for spelling or grammar errors.

Scribe's Name ______________________________ Date ____________________

32. Eddie (or Edie) Electron *(continued)*

In-Class Group Assignment

1. Pick up your writing papers from your teacher and form groups as instructed by your teacher.
2. Listen carefully while one person reads her story out loud to the group.
3. Stop the reader as he or she covers one of the required terms in the assignment. Mark a star (*) on the paper at the place where that part of the assignment is explained.
4. After each person finishes reading aloud, the reader should list at the top of the front page the number of stars marked on his or her paper. Then take about five minutes to write your reaction to their story. You must include one good thing in or about the story and one improvement you might make in the story.
5. Give your reaction paper to the writer of the story.
6. Continue to listen, stop, and react to each member of the group.
7. Select one member of the group to be a scribe.
8. Write each group member's name on a sheet of notebook paper or on the sheet your teacher provides.
9. Write a group consensus answer to each of the questions.
   a. List three *specific* materials that are good **conductors**. (For example, "metals" is too general an answer.)
   b. List three materials that are good **insulators**.
   c. Batteries come in several different types. Name three.
   d. Why does a light bulb glow?
   e. Why is it not a good idea to connect several thin wire extension cords end-to-end when you need a long extension cord?
10. Staple the stories underneath the consensus answers in the order that the group members' names are listed on the answer sheet.
11. Turn the packet in to your teacher.

Scribe's Name ______________________________ Date ____________________

33. Fish Respiration: This Lab's All Wet!

In this lab you will be recording the respiration rate of a goldfish under a particular set of conditions. You will also be comparing your data with the data of other pairs of people in your class.

Procedures

1. Get into groups as directed by your teacher.
2. Listen to your teacher to find out which of three groups you and your partner(s) are assigned to: **room temperature, increased temperature, decreased temperature**.
3. Put the appropriate amount of water in your beaker.
4. Record the temperature of the water.
5. Add a goldfish to the beaker and the water.
6. Record the number of gill or mouth movements your fish performs in one minute.

AT NO TIME DURING THIS LABORATORY SHOULD THE WATER TEMPERATURE IN YOUR BEAKER GO ABOVE 40°C OR BELOW 10°C.

IF YOUR FISH DIES, YOUR GROUP GETS "O" (zero).

7. Add **ice**, **warm water**, or **nothing** to your beaker depending on your assigned group.
8. Wait one minute.
9. Take the temperature of the water and record it.
10. Record the number of gill or mouth movements your fish performs in one minute.
11. Wait five minutes.
12. Take the temperature of the water and record it.

(continued)

Scribe's Name ______________________________ Date ____________________

33. Fish Respiration: This Lab's All Wet! *(continued)*

13. Record the number of gill or mouth movements your fish performs in one minute.

14. Wait five minutes.

15. Take the temperature of the water and record it.

16. Record the number of gill or mouth movements your fish performs in one minute.

17. Wait five minutes.

18. Take the temperature of the water and record it.

19. Record the number of gill or mouth movements your fish performs in one minute.

20. Now copy the data from two other groups who each did a different temperature than you did. You should end up with data for warm, cold, and room temperature groups.

Scribe's Name ______________________ Date ______________

33. Fish Respiration: This Lab's All Wet! *(continued)*

At-Home Assignments

1. As a part of your lab write-up, you will make a graph of all three groups of data. This graph should have three colors. The horizontal axis should be the temperature in degrees C. The vertical axis should be number of gill (or mouth) movements per minute.

2. Make a general statement relating the temperature to the **respiration** in fish.

3. What is a *hard scientific word* that is used to describe a condition in fish that could account for the results you achieved? You'd better explain this a little.

4. Do you think that a snake would have a similar change in rate of respiration depending on the temperature? What about a human being? Why?

5. What is a major flaw in this laboratory procedure?

Scribe's Name ______________________________ Date ____________________

34. Move Over, Bill Watterson

The task in this exercise is to devise a cartoon on a topic in physical science. You may choose to do a four-frame cartoon (like *Calvin and Hobbes* or *Garfield*) or a single-frame cartoon (like *The Far Side*). A third option is to do a political cartoon on some aspect of physical science. Check with your teacher to see if there is a list of acceptable topic areas for you to choose from before you begin your cartoon.

1. You may work on your cartoon by yourself or with a single partner.
2. The cartoon should be drawn, and colored, on a single sheet of unlined typing paper. All words and pictures should be dark or brightly colored.
3. Hopefully, no explanation of your cartoon will be necessary.
4. The cartoon must be accurate in its portrayal of the physical science you use. In other words, you cannot have a magnet attracting wood unless you correct that fallacy in the cartoon, etc.
5. You will have twenty-five minutes to complete this task. At the end of the time allowed for production, an opaque projector will show your cartoon to the entire class on the screen. Therefore, neatness counts.

Good Humor!